Journeyman Plumber's
Exam
SECRETS

Study Guide
Your Key to Exam Success

Plumber's Test Review for the
Journeyman Plumber's Exam

Dear Future Exam Success Story:

Congratulations on your purchase of our study guide. Our goal in writing our study guide was to cover the content on the test, as well as provide insight into typical test taking mistakes and how to overcome them.

Standardized tests are a key component of being successful, which only increases the importance of doing well in the high-pressure high-stakes environment of test day. How well you do on this test will have a significant impact on your future, and we have the research and practical advice to help you execute on test day.

The product you're reading now is designed to exploit weaknesses in the test itself, and help you avoid the most common errors test takers frequently make.

How to use this study guide

We don't want to waste your time. Our study guide is fast-paced and fluff-free. We suggest going through it a number of times, as repetition is an important part of learning new information and concepts.

First, read through the study guide completely to get a feel for the content and organization. Read the general success strategies first, and then proceed to the content sections. Each tip has been carefully selected for its effectiveness.

Second, read through the study guide again, and take notes in the margins and highlight those sections where you may have a particular weakness.

Finally, bring the manual with you on test day and study it before the exam begins.

Your success is our success

We would be delighted to hear about your success. Send us an email and tell us your story. Thanks for your business and we wish you continued success.

Sincerely,

Mometrix Test Preparation Team

Need more help? Check out our flashcards at:
http://MometrixFlashcards.com/Plumber

TABLE OF CONTENTS

Top 20 Test Taking Tips

1. Carefully follow all the test registration procedures
2. Know the test directions, duration, topics, question types, how many questions
3. Setup a flexible study schedule at least 3-4 weeks before test day
4. Study during the time of day you are most alert, relaxed, and stress free
5. Maximize your learning style; visual learner use visual study aids, auditory learner use auditory study aids
6. Focus on your weakest knowledge base
7. Find a study partner to review with and help clarify questions
8. Practice, practice, practice
9. Get a good night's sleep; don't try to cram the night before the test
10. Eat a well balanced meal
11. Know the exact physical location of the testing site; drive the route to the site prior to test day
12. Bring a set of ear plugs; the testing center could be noisy
13. Wear comfortable, loose fitting, layered clothing to the testing center; prepare for it to be either cold or hot during the test
14. Bring at least 2 current forms of ID to the testing center
15. Arrive to the test early; be prepared to wait and be patient
16. Eliminate the obviously wrong answer choices, then guess the first remaining choice
17. Pace yourself; don't rush, but keep working and move on if you get stuck
18. Maintain a positive attitude even if the test is going poorly
19. Keep your first answer unless you are positive it is wrong
20. Check your work, don't make a careless mistake

Comply with General Laws

Plumbing permits

Before starting any plumbing job, it is necessary to make sure that proper permits are obtained. Most municipalities require plumbers to obtain permits in order to keep records of all plumbing work performed in the municipality. Usually, the plumbing contractor is responsible for obtaining permits. The plumbing contractor is required to submit the blueprints, specifications for the job, and the proposal for the plumbing plan so that the plumbing inspector can check the documents for compliance with appropriate codes and regulations. If the documents are approved, the municipality issues plumbing permits for the work to be performed. Plumbing permits are required for such work as:

- water service connection
- building sewer connection
- installation of plumbing systems and fixtures in the building

The plumbing contractor is expected to pay a fee for each permit.

License requirements

The following are the license requirements for plumbing specialists:

- Journeyman Licenses are required of professionals who install, maintain, alter, or repair plumbing fixtures and systems under the supervision of a Journeyman Plumber
- Journeyman Plumber Licenses are required of persons who perform or contract to perform plumbing services
- Class I Journeyman Plumber Licenses are restricted to plumbing involving single family dwellings, one-level dwellings designed for no more than two families, and commercial structures not exceeding 10,000 square feet
- Class II Journeyman Plumber Licenses are unrestricted
- Nobody should engage in the plumbing business as a journeyman plumber unless this person has a valid license
- Nobody should engage in the plumbing business as a Journeyman Plumber unless this person has a valid license
- No partnership or corporation should engage in the plumbing business unless its employees and partners have valid licenses for Journeyman Plumbers

Journeyman Plumber and Journeyman Plumber license applicants
Journeyman Plumber requirements are as follows:
- Applicants for statewide Journeyman Plumber license must present proof of a minimum of 3 years experience in plumbing work as covered by regulations of the State Plumbing Code
- Applicants for statewide Journeyman Plumber license must present proof of a minimum of 5 years experience in plumbing work as covered by regulations of the State Plumbing Code. At least 2 years of experience must be Primary Experience, such as installation of plumbing pipes and fixtures by a plumbing contractor, Journeyman Plumber, journeyman plumber, plumbing foreman, plumbing superintendent, a military plumber with a rank of at least a 3rd class petty officer; mechanical engineer with responsibility for follow-up project inspection.
- Applicants for Class II Unrestricted Journeyman Plumber license must document experience in commercial or industrial plumbing.

Complaint procedures

The following are the Construction Industry Licensing Board complaint procedures:
- A person wishing to file a complaint concerning a violation of the licensure requirements should submit the complaint in writing to the Construction Industry Licensing Board
- A board investigator will contact the person who filed the complaint and investigate the complaint promptly
- If it becomes known to the Board that a person practices a regulated business or profession without a license, the Board will issue a cease and desist order, a fine, and transfer the case to local authorities for criminal prosecution
- If a licensee has violated statute or board rule, the Board will impose a disciplinary action, such as a fine, reprimand, suspension or revocation of the license. The board cannot order a licensee to complete or correct work that led to the complaint

License reinstatement

The following is the procedure for license reinstatement:
- An applicant should complete the application form, sign and mail the form attaching documentation of completion of continuing education and the fee made payable to the "State Construction Industry Licensing Board"
- An applicant's license number should be recorded on the check or money order
- An applicant may submit a name change with the reinstatement application
- A name change must be submitted in writing, accompanied by supporting legal documentation (i.e., copy of marriage license, divorce decree, court contractor.)

- If an applicant has a conviction or board disciplinary action, a certified court record or board disciplinary order must be attached
- License holders are required to report to the board any felony or drug related conviction within 10 days of the date of the conviction

Renewal procedures

The following are the plumbing license renewal procedures:
- To qualify for renewal, conditioned air contractors, electrical contractors, and plumbers must have completed continuing education
- A plumbing specialist must apply for renewal at least six weeks prior to the expiration date
- A renewal form is mailed to all current licensees approximately six weeks prior to renewal
- If a licensee needs a change of address service, he/she should advise the Board office in writing

The renewal fee may vary by state and is usually seventy-five dollars for a two-year period, except for the thirty-five dollar renewal fee for journeyman plumber and utility foreman licenses. The renewal fee is payable to the State Construction Industry Licensing Board.

Refusing or revoking licenses

The following are the grounds for refusing to grant or revoking plumbing licenses:
- Failure to demonstrate the qualifications or standards for a license contained in the State Code, or under the laws, rules, or regulations under which licensure is sought or held
- Knowingly making misleading, deceptive, untrue, or fraudulent representations in the practice of a business or profession where licensure is sought or held
- Conviction of any felony or any crime involving moral corruption
- Revoked, suspended, or annulled license by any lawful licensing authority other than the board
- Engagement in any unprofessional, immoral, unethical, deceptive conduct or practice harmful to the public, when conduct or practice materially affects the fitness of the licensee or applicant to practice a business or profession licensed
- Knowingly assisting in any way to any unlicensed person or any licensee whose license has been suspended or revoked

Pre-excavation requirements

The excavation operator is responsible for the following according to the State Law pre-excavation requirements:
- Submitting a new locate request if additional areas outside the scope of the original ticket need to be marked
- Renewing a locate request if excavation does not start within 10 working days
- Renewing a locate request when markings are no longer visible and excavation is still continuing
- Renewing a locate request if excavation will continue past 20 working days
- Using an excavation method approved by the utility owner or hand-dig within 18 inches either side of the markings (safety zone)
- Notifying the utility owner of damage immediately. If the damage permits the escape of hazardous gas or liquid, the police and fire departments must also be notified

Temporary underground facility markers

Temporary underground facility markers should be made in paint, chalk, flags, stakes, or any combination of these, and should conform to the standards of the American Public Works Association Uniform Color Code. The following are the standards:
- Electric power distribution and transmission – Safety red
- Municipal electric systems – Safety red
- Gas distribution and transmission – High visibility safety yellow
- Oil distribution and transmission – High visibility safety yellow
- Dangerous materials, product lines – High visibility safety yellow
- Telecommunications systems and cable television – Safety alert orange
- Temporary survey markings – Safety pink
- Police and fire communications – Safety alert orange
- Water systems – Safety precaution blue
- Sewer and storm drainage systems – Safety green
- Proposed excavation or construction boundaries – White
- Reclaimed water, slurry, and irrigation facilities – Purple

ADA

Americans with Disabilities Act (ADA) provides comprehensive civil rights protection to individuals with disabilities in the areas of employment, public accommodations, state and local government services, and telecommunications. Title II of the ADA requires state and local authorities to ensure equal opportunity for qualified individuals with physical and mental disabilities.

The following are the main provisions:
- General Nondiscrimination Requirements – prohibits the denial of services or benefits on the basis of disability
- Employment – prohibits discrimination in employment in programs or activities that receive Federal financial assistance
- Program Accessibility – prohibits denial of benefits of a public entity programs, activities, and services to individuals with disabilities because its facilities are inaccessible
- Equally Effective Communications – requires a public entity to ensure that its communications with individuals with disabilities are as effective as communications with others

Underground facility location identification

The approximate location of underground facilities is defined as a strip of land at least 3' wide.

Requirements
Before excavation, State Excavation Authorities should notify all operators of underground facilities in the affected area of the proposed activity within four working hours after receiving notification of intent to excavate.

The operator should identify the approximate location of the facilities by field-marking on the surface by paint, dye, stakes, or other visible marking.

When an underground facility is located, the operator should identify the approximate center line, estimated depth, and dimensions of the underground facility

When excavating within the approximate location of an underground facility, the excavator should uncover the facility using a method approved by the operator.

The operator and excavator should establish and maintain coordination regarding location, marking, and identification of the facilities until excavation is completed.

ANSI

The American National Standards Institute (ANSI) is a regulatory non-profit organization that coordinates development and use of the United States standards. ANSI represents the needs and views of U.S. stakeholders in standardization forums around the world. The organization supervises creation, promotion, and use of norms and guidelines that have a direct affect on such businesses as production of acoustical devices, construction equipment, dairy and livestock, energy distribution, public entity accessibility, etc. ANSI has an authority to accredit programs that evaluate worldwide conformance to standards. These programs include the ISO

9000 (quality) and ISO 14000 (environmental) management systems. In addition, ANSI facilitates development of American National Standards (ANS) by accrediting procedures of standards developing organizations (SDOs).

<u>Public facilities</u>
The following are the ADA requirements for new constructions and alterations:
- If there is more than one drinking fountain on a floor, 50 percent of all the fountains should be accessible to persons using wheelchairs
- If there is only one drinking fountain on a floor, it must be accessible both to individuals who use wheelchairs and to individuals who have trouble bending or stooping
- Multifamily dwellings constructed after March 31, 1993, with a building entrance on an accessible route should be designed with reinforcements in bathroom walls to allow later installation of grab bars around the toilet, tub, shower stall, and shower seat
- Multifamily dwelling units should contain kitchens and bathrooms where an individual in a wheelchair can maneuver about the space

<u>Public restrooms</u>
The following are the ADA requirements for new constructions and alterations:
- Accessible toilet rooms, bathrooms, bathing facilities, and shower rooms should be provided in a reasonable number (ANSI 4.22). The reasonable number means that on each floor which is to be made accessible to and usable by handicapped persons with disabilities there should be at least one toilet room, bathroom, bathing facility, and shower room at a reasonable location
- Every public and common use bathroom must be accessible. At least one stall must be accessible. The stall should be 5 by 5 feet. If there are six or more stalls, there must be one accessible stall and one three feet wide stall

State Water Conservation laws

State Water Conservation laws are created by such organizations as State Environmental Conservation Committees, Departments of Soil and Water Conservation, etc. These organizations ensure that public and private entities comply with the environmental standards accepted in the state.

The water conservation laws are reflected in water conservation plans which cover such areas as:
- requirements for wholesale public water suppliers
- requirements for municipal water use by public water suppliers
- requirements for industrial water conservation
- requirements for agricultural water uses
- provisions of water-saving plumbing fixture programs
- provisions of The Environmental Performance Standards for Plumbing Fixtures. This law requires that only plumbing fixtures that conform to specific water use efficiency standards can be sold in the state

Property owners

The following are the state water conservation laws concerning propriety owners:
- The power of homeowner's associations to discourage outdoor water conservation should be limited. Many property owners' associations have rules that require certain amounts and types of turf grass coverage; promote excessive maintenance standards and irrigation systems; and prohibit native or climatically appropriate landscapes and rainwater harvesting systems. These rules undermine water conservation goals
- The rules should be developed to allow for increased ability to use gray water in place of potable water for common irrigation purposes. Property owners should be able to install gray water systems to conserve water and use such water resources as wastewater from clothes washing machines, showers, bathtubs, and hand washing sinks

Energy Policy Act

The Energy Policy Act was accepted in 1992. The Act was designed to establish water conservation standards for the manufacture of plumbing fixtures. Four types of plumbing fixtures are covered by the Act: toilets, kitchen and lavatory faucets, showerheads, and urinals. The standards apply to all models of the fixtures manufactured after January 1, 1994 with a few exceptions. For each model of a regulated plumbing fixture, manufacturers must submit a compliance statement to the Department of Energy to certify that the model meets the water conservation standards and that all required testing has been conducted according to the test requirements prescribed in the regulations. In addition, manufacturers are prohibited by the Department's regulations from distributing any fixture that does not meet the water conservation standard prescribed under the Energy Policy Act. The violation of the Act results in a civil penalty of not more than $110 per violation.

Maximum water usage

Fixture Type	Average Water Consumption
Water Closets - low profile one-piece gravity flush type or handicap accessible floor mount, gravity flush closets (minimum 17" height)	1.6 gallons per flush
Water Closets - flushometer tank of close-coupled two-piece gravity flush type	1.6 gallons per flush
Water Closets – flushometer valve, floor mount	2.5 gallons per flush
Water Closets - wall mount	3.5 gallons per flush
Urinals	1.0 gallons per flush
Public Lavatory Faucets	0.5 gpm
Showerheads	2.5 gpm

Uniform Plumbing Code

The following are the Uniform Plumbing Code requirements for materials:
- All materials used in a plumbing system and procedures for their installation must be approved by the code enforcement officer who has the authority to change the provisions of the local code if there are no health, safety, and public welfare risks
- A property owner may apply for a variance from the standard code requirements for materials provided there are certain conditions that warrant a hardship
- The use of previously used materials can be challenged by the local code officer. To continue using the same materials, it is necessary to test them and prove their working condition
- The local code enforcement officer has the authority to allow the use of alternative materials if their safety, durability, and quality have been proved to be at least equal to the standard approved materials

Comply with Regulations

Standard Gas Code

According to the Standard Gas Code requirements for connecting concealed indoor gas pipes the following types of connections should not be used:
- unions
- tubing fitting
- right and left couplings
- bushing
- swing joints
- compression couplings made by combining several fittings

Instead, the following connections should be used:
- Pipe fittings, such as elbows, tees, and couplings
- Tubing joint by brazing
- Fittings than can sustain such forces as temperature expansion or contraction, vibration, fatigue due to the geographic location, application, or operation
- Welding, flanges, or ground joint union should be used to reconnect the pipes when it is necessary to insert fittings in a previously installed concealed gas pipe

Standard Building Code Congress Gas Code

The following are the general requirements for gas systems and connections as prescribed by the Standard Building Code Congress Gas Code:
- The storage system for liquefied petroleum gas should be designed and installed in accordance with NFPA 58
- Modifications and additions to the existing gas piping systems should be performed in accordance with the Gas Code
- The gas piping system should not be interconnected on the outlet side of the meters when two or more meters are installed in the same premises but supply different users
- An approved backflow prevention device should be installed when a supplemental gas supply for standby use is connected downstream from a meter

National Fuel Gas Code

The following are the National Fuel Gas Code requirements for plastic pipes:
- Plastic pipes should be installed underground outdoors only. Plastic pipes can be installed aboveground only if an anodeless riser is used; or, if a wall head adapter is used, and plastic pipes are inserted in piping made with material permitted for indoor use

Connections between outdoor underground metallic and plastic pipes can be made with either of the following:
- ASTM D 2513 Standard Specification for Thermoplastic Gas Pressure Pipe, Tubing, Fitting, Category 1 transition fittings
- ASTM F 1973 Standard Specification for Factory Assembled Anodeless Risers and Transition Fittings In Polyethylene and Polyamide 11 Fuel Gas Distribution Systems

Code of Federal Regulations

Title 29, Part 1926
Part 1926 of Title 29 of Code of Federal Regulations was designed to promote the safety and health standards promulgated by Section 107 of the Contract Work Hours and Safety Standards Act. Section 107 requires that the following conditions should be observed in each contract for construction, alteration, and/or repair, including painting and decorating entered into under legislation subject to Reorganization Plan: No contractor or subcontractor should require any laborer or mechanic employed by the contractor or subcontractor to work under unsanitary, hazardous, or dangerous to his/her health or safety working conditions. These conditions are determined under construction safety and health standards promulgated by the Secretary of Labor by regulation.

Subpart C of Part 1926, Title 29, Code of Federal Regulations requires the following of contractors:
- Contractors or subcontractors should provide sanitary and safe working conditions for laborers and mechanics they legally employ
- Employers are responsible to provide programs that prevent on-the-job accidents
- These programs should be made available for regular inspections of the job sites, materials, and equipment by competent specialists designated by the employers
- Employers should not require laborers and mechanics to use materials, tools, and equipment that do not comply with the provisions of Subsection C
- Employers should only employ those employees who are fully qualified for the task to be performed

Federal-Aid Highway Act

The Federal-Aid Highway Act, or Title 23 of the US Code, promulgates that the Secretary of Labor should ensure that all laborers and mechanics employed by contractors or subcontractors on the construction work performed on highway projects on the Federal-aid highways should be paid wages at specific rates. These rates should be no less than those provided on the same type of work on similar construction in the immediate locations. The Secretary of Labor should consult with the highway department of the State where a project on any of the Federal-Aid systems is going to be performed before determining the minimum wage. The Act is not applicable to apprenticeship and skill training programs certified by the Secretary of Transportation as promoting equal employment opportunity in connection with Federal-Aid highway construction programs.

First aid OSHA provisions

The following are the OSHA provisions for first aid:
- The employer should make available first aid services and provisions for medical care for each legally employed employee
- The employer should ensure the availability of medical personnel for advice and consultation on matters of occupational health
- If professional medical assistance cannot be provided, a person with a valid certificate in first-aid training from the U.S. Bureau of Mines, the American Red Cross, or equivalent training should be available at the worksite
- First aid supplies should be easily accessible
- If 911 services are not available, the phone numbers of physicians, hospitals, and ambulances should be made available
- In areas where the eyes or body of any person may be exposed to harmful corrosive materials, emergency showers and eye wash fountains should be provided within the work area

Housekeeping OSHA requirements

The following are the OSHA requirements for housekeeping:
- During the construction, alteration, and repairs, lumber with protruding nails and other debris should not be kept close to work areas, passageways, and stairs
- Flammable scrap and debris should be regularly disposed of during the course of construction. Safe means of disposal should be provided
- Containers should be provided for the collection and separation of waste, trash, oily and used rags, and other refuse
- Containers used for garbage and other oily, flammable, or hazardous wastes, such as caustics, acids, harmful dusts, etc. should be equipped with covers
- Garbage and other waste should be removed frequently and regularly

Safety training and education

The following are the employer's responsibilities in safety training and education:
- The employer should be available for the safety and health training programs provided by the Secretary of Labor
- The employer should instruct employees on how to recognize and avoid unsafe conditions and how to control or eliminate any hazards
- The employer should instruct employees required to handle or use harmful substances, plants, or animals on the rules of safe handling and use, personal hygiene, and personal protective measures
- The employer should instruct employees required to handle or use flammable liquids, gases, or toxic materials on the safe handling and use of these materials
- The employer should instruct employees required to enter into confined or enclosed spaces on the nature of the hazards involved, the necessary precautions to be taken, and the use of protective and emergency equipment

State boiler code

Boiler codes differ by state; however, most of them require the following:
- The construction of boilers and pressure vessels and their installation should be in accordance with minimum requirements for safety from structural and mechanical failure and excessive pressures
- Boilers and pressure vessels should comply with Sections I, III, IV, VIII, X, and PVHO-1 of the American Society of Mechanical Engineers' (ASME) Boiler and Pressure Vessel Code and the American National Standards Institute (ANSI) B31.1.0 Power Piping Code
- All fossil fuel fired boiler installations with fuel input ratings of less than 12,500,000 Btu/hr should comply with the fuel train requirements of ASME CSD-1-2002, Controls and Safety Devices for Automatically Fired Boilers (CSD-1)

Boiler registration and installation permit

All boilers and pressure vessels should be registered with the National Board of Boiler and Pressure Vessel Inspectors to ensure safety with the exception of cast-iron boilers and pressure vessels bearing the ASME "UM" stamp which do not need to be registered. A permit must be obtained from the Director prior to the following:
- Installation or replacement of new or used boilers and pressure vessels
- Installation of rental boilers
- Certification of boilers as "automatic"
- Certification of boilers as "monitored"

- Alteration or modification of existing control systems on boilers certified as "automatic" or "monitored"
- Replacement or modification of fuel burners, changing fuels, or adding different fuel combinations

Inspection requirements

Boiler and pressure vessel systems requiring permits should be inspected by the Director. The permit applicant should make the boiler and pressure vessel systems accessible and exposed for inspection purposes. Inspectors are not liable for expenses required to remove or replace any material required for inspection. A final inspection should be made upon completion of the installation of a boiler and pressure vessel system. Until authorized by the Director, boiler and pressure vessel systems should not be connected to the energy fuel-supply lines. The Director may require a re-inspection in the following cases:
- work is not complete
- corrections are not made
- inspection record is not properly posted on the work site
- approved plans are not available to the inspector
- deviations from plans have been made without proper approval
- failure to provide access on the date of the inspection

Insurance company representatives
Authorized insurance company representatives may inspect boilers and pressure vessels. The following are requirements for such inspections:
- Inspections should be conducted by persons holding an active commission from the National Board of Boiler and Pressure Vessel Inspectors
- Insurance companies must annually notify the Director in writing of inspectors who will be conducting inspections
- Authorized inspectors should report to the Director on official forms prescribed by the Director
- Insurance company inspectors should immediately notify the Director of any suspension of insurance coverage due to dangerous conditions
- Authorized insurance companies providing insurance coverage of jurisdictional objects should notify the office of the State Boiler Code within 30 days for any new insurance in effect or any discontinuance of insurance coverage of jurisdictional objects

Trenching safety procedures

Trenching safety procedures slightly differ from state to state. The following procedures are common for most of the states:
- Locate all underground utilities prior to digging
- Analyze soil to determine soil type. If not sure, assume it is Type C. Slope trench sides appropriate to the type of soil, or provide shoring or trench box
- Increase slope of trenches exposed to vibrations of construction equipment, operations, traffic
- Keep excavated materials at least 2 feet away from the edge
- Don't allow water to accumulate in the trench
- Keep heavy loads away from the trench
- Provide ladder, steps, or ramp within 25 feet of travel from anywhere in the trench
- Use professional engineering for trenches 20' or deeper
- An authorized inspector, formally trained in Trenching Safety, must be present at all times during trench work

Installation permit application

To obtain a permit, the applicant should file an official application in writing on a form provided by the Director. Each application should provide the following information:
- Identification and description of the work to be covered by the permit
- Description of the land on which the proposed work is to be done; for example, legal description, property address, or other description that will identify and definitely locate the proposed building or work
- Plans and/or specifications in the standard ASME form (Manufacturers Data Report)
- The signature of the owner of the property or building, or authorized agent, who may be required to submit evidence to indicate such authority
- The name of the owner and contractor and the name, address, and phone number of a contact person

Permit issuance procedure

The Director of the State Boiler Code and other departments of the City review the application, plans, and specifications filed by a permit applicant. If the description of work in the application conforms to the requirements of the State Boiler Code and other related regulations, the Director issues a permit after the permit fees have been paid.

A permit issued by the Director expires in 18 months from the date of issuance. The Director may determine if a permit should be issued for a shorter period. In this case, a permit will expire in less than 18 months. A permit may be renewed once, if the following conditions are met:

- Application for renewal is made within the 30 day period immediately preceding the date of expiration of the permit
- The work authorized by the permit has been started and is progressing at a rate approved by the Director

Internal inspections

The owner or user should prepare a boiler or pressure vessel for internal inspection by the Director or insurance company. The following is a typical preparation for internal inspection for boilers:

- Water should be drawn off. The boiler should be washed
- Manhole and handhole plates, wash-out plugs, and water column connections should be removed
- The furnace and combustion chambers should be cooled and cleaned
- All grates of internally fired boilers should be removed
- Brickwork and/or refractory should be removed in order to determine the condition of the boiler headers, furnace, supports, and other parts
- The low water cutout should be disassembled to a degree required by the inspector

Pressure and temperature relief requirements

The following are the requirements for pressure and temperature relief controls:

- The discharge from liquid relief valves should be piped to within 18 inches of the floor or to an open receptacle
- When the operating temperature is more than 140°F, liquid relief valves should be equipped with a means of tempering and cooling the discharge prior to entering the drainage system
- Safety valve discharge from boilers and pressure vessels containing steam should be directed upward to a minimum of 6 feet above the boiler room floor or horizontally to an inaccessible area of the boiler room
- When the discharge from safety valves would result in a hazardous discharge of steam inside the boiler room, or when the discharge of multiple safety valves on boilers exceeds the capacity of 1,000 pounds of steam per hour, it should be extended outside the boiler room to a safe location

Controls safety requirements

The following are the safety requirements for controls:
- Required electrical, mechanical, safety and operating controls should be approved by an authorized testing agency
- Fuel burners should be listed by a nationally recognized testing agency. Burners which are integral parts of boilers should be listed as part of the overall boiler-burner assembly
- Burners installed after June 1, 1987 which are capable of burning two or more fuels should be equipped with a fuel selector switch designed and constructed to prevent switching from one fuel to a different fuel without a physical stop in the center/off position
- Steam boilers should be provided with a pressure gauge and a water level glass. Water boilers should be provided with a pressure gauge and a temperature indicator. Hot water supply and storage tanks should be provided with a pressure gauge and temperature gauge. Gauges should be kept in good working condition

Plumbing Divisions of State Health Departments

The following are the services offered by Plumbing Divisions of State Health Departments:
- Inspection of plumbing in all new private homes and commercial dwellings to determine compliance with state laws and rules
- Collection of water samples for bacteriological analysis
- Inspection of the plumbing in all existing homes prior to hook-up to the sanitary sewer lines
- Facilitation of disinfection procedures for drinking water wells
- Issuance of well permits
- Issuance of permits to use existing sewer systems
- Installation of sewage disposal systems (septic tanks)
- Processing of septic installer registration
- Sewage evaluation
- Processing and issuance of plumber registrations and permits
- Installation of semipublic wastewater treatment systems
- Inspection of public and semi-public water supply systems

State Boards of Health

State Boards of Health have the authority to adopt rules and regulations for State Plumbing Advisory Boards. State Boards of Health direct State Health Departments to propose rules, regulations, standards, policies, and procedures to secure the intent and to designate alternative minimum standards due to local climatic or other appropriate conditions. Such rules, regulations, standards, policies, or procedures do not have the effect of waiving structural or fire performance requirements specifically provided for in state codes. Rules, regulations, standards, policies, or procedures proposed should not violate accepted engineering practices involving public safety. State Boards of Health may direct the Plumbing Advisory Board to review and comment upon rules, regulations, standards, policies, and procedures proposed by State Health Departments.

Piping materials requirements

The following materials should comply with these standard requirements:
- Acrylonitrile butadiene styrene (ABS) plastic – ASTM D 2468
- Chlorinated polyvinyl chloride (CPVC) plastic – ASTM F 437; ASTM F 438; ASTM F 439
- Polyethylene plastic – ASTM D 2609
- Polyvinyl chloride (PVC) plastic – ASTM D 2464; ASTM D 2466; ASTM D 2467; CSA CAN/CSA-B137.2
- Copper or copper alloy – ASME B 16.15; ASME B 16.18; ASME B 16.22; ASME B 16.23; ASME B 16.26; ASME B 16.29; ASME B 16.32
- Metal insert fittings with copper crimp ring SDR9 (PEX) tubing – ASTM F 1807
- Steel – ASME B 16.9; ASME B 16.11; ASME B 16.28
- Cast iron – ASME B 16.4; ASME B 16.2
- Gray iron and ductile iron – AWWA C110; AWWA C153
- Malleable iron – ASME B 16.3

Measures against water contamination

State Departments of Health require plumbing installations to be equipped with backflow prevention systems in order to avoid contamination of the public water supply system with the sewage water from private drainages. The backflow prevention can be facilitated by a variety of devices which should be installed immediately downstream of the water main. These devices include reduced pressure backflow preventers, double check valve assemblies, dual check valve assemblies, atmospheric vacuum breakers, hose connection vacuum breakers, etc. Another method of backflow prevention recommended by the State Departments of Health is installation of air gaps between the water outlet and the flood level rim of the fixture. It should be noted, however, that the abovementioned backflow prevention methods do not protect from contamination inside the building.

Joint material weight and size

The following are the recommendations for joint material weight and size:
- Brass caulking ferules should be of red brass or heavy cast red brass with weights

The table below illustrates the recommended weights and sizes for brass caulking ferules.

Pipe Size	Inside Diameter	Length	Weight
2 inches	2¼ inches	4½ inches	1 lb. – 0 oz.
3 inches	3¼ inches	4½ inches	1 lb. – 12 oz.
4 inches	4¼ inches	4½ inches	2 lb. – 8 oz.

Soldering bushings should be of red brass pipe or of heavy cast red brass.

The table below illustrates the recommended weights and sizes for brass soldering bushings.

Pipe Size	Minimum Weight	Pipe Size	Minimum Weight
1¼ inch	6 oz.	2½ inches	1 lb. – 6 oz.
1½ inch	8 oz.	3 inches	2 lb. – 0 oz.
2 inches	14 oz.	4 inches	3 lb. – 8 oz.

Pipe thickness recommendations

The following are the recommendations for pipe thickness:
- Sheet lead should weigh no less than 4 pounds per square foot; lead bends and traps should be no less than ⅛ inch thick
- Sheet copper used for safe pans should weigh no less than 12 ounces per square foot; sheet copper used for vent terminal flashings should weigh no less than 8 ounces per square foot
- Sheet iron used for pipe duct should have wall thickness of 26 for 2- to 12-inch pipe; 24 for 13- to 20-inch pipe; 22 for 21- to 26-inch pipe
- Thickness of floor- and wall-flanges for water closet should be no less than 3/16 inch for cast iron, and no less than ⅛ inch for brass

Work Planning/Organizing

Written specifications

Written specifications supplement the blueprints by giving instructions and explanations that cannot be effectively expressed in the blueprint. For example, the blueprint may indicate the location of water closets, bathtubs, lavatories, etc., while the written specification may describe the required quality, manufacturer, color, and style of these fixtures. Specifications explain what type of piping should be used for each plumbing system. They also include information on legal responsibilities, insurance, quality of workmanship. Usually, specifications are filled in a form provided for home owners by local authorities and contain such information as fixture quantity, size, color, make, identification number; material, weight, size, and shape of piping; and details of any special equipment to be installed, for example, dishwashers, garbage disposal units, etc.

Bill of materials

A bill of materials is a list of all the materials, fixtures, and equipment necessary to complete a plumbing project. The bill of materials is based on the interpretation of blueprints and specifications and consists of two parts: a job takeoff and an estimate. A takeoff includes a count and check-off of all the plumbing materials, such as piping, tubing, fittings, fixtures, boilers, heaters, etc. required for installation as specified on the blueprint. An estimate includes materials not specified on the blueprints; for example, pipe hangers, nails, pipe joint compound, oakum, lead, etc. The job takeoff and the estimate are combined to document a complete fill of materials for a specific plumbing system.

Blueprints

A blueprint is a working drawing that delivers information on how the architect and the electrical, mechanical, and structural engineers involved in the building construction want the various parts of the plumbing, heating, and electrical systems laid out.

Large-scale construction projects usually have different types of blueprints separated into sets. In small-scale construction projects, structural and mechanical blueprints may be incorporated in the architectural blueprint.

Structural blueprints
Structural blueprints represent the supporting structure of the building. The structural blueprints include pilings, footings, foundation walls, columns, beams, floor slabs, and roofing.

Architectural blueprints

Architectural blueprints are a representation of the complete building plan without structural and mechanical details. Architectural blueprints depict framing, walls, partitions, wall finish schedules, trim, cabinets, and all measurements for all the parts of the plan.

Mechanical blueprints

Mechanical blueprints represent the plumbing, heating, and electrical systems giving a complete drawing of all the fixtures and appliances to be installed according to the architectural plan.

⊸(M)⊸	WATER METER	— - — - —	COLD WATER
— - - —	HOT WATER	- - - - - - - - - -	VENT LINE
———	SANITARY WASTE	— G —	GAS PIPE
⊸▷◁⊸	GATE VALVE	⊸▷◁⊸	WATER HEATER SHUT OFF
(WC)	WATER CLOSET	(LAV)	LAVATORY
(WH)	WATER HEATER	DW	DISHWASHER
CW	CLOTHES WASHER	⊘	FLOOR DRAIN
CLEAN OUT	CLEAN OUT	VTR	VENT THRU ROOF
90° ELBOW	90° ELBOW	O⊸	PIPE TURNS UP
⊝	PIPE TURNS DOWN	╫	TEE
╫	UNION	⊤	CAP

Plumbing takeoff sheet

The following should be included in the plumbing takeoff sheet:

- Fixture takeoff requirements, which are listed on the drawings by name, stock number, and quantity, are determined after careful study of the drawings and specifications
- Pipes are listed on the takeoff sheet by size, material type, stock number, and lineal feet required. For example, ½" copper tubing – stock number – 40 lf
- Fittings are listed by size, material, classification, stock number, and required quantity. For example, ½" galvanized cast iron elbows 90° – stock number – 30 ea
- Soil pipe is listed by size, material, and strength or weight. Fittings for soil pipe are listed by size, material, classification, strength or weight, and quantity. For example, 4" quarter-bend extra heavy cast iron – 4 ea

- 21 -

Takeoff forms

The following are takeoff forms:
- Plumbing fixture quantity sheet – documents the number of plumbing fixtures by floor. The "Total" column indicates the total amount of fixtures in the entire building
- Equipment quantity sheet – documents plumbing equipment from each system; includes such specifications as capacity, description, etc.
- Piping quantity sheet – documents the diameter and the linear feet of piping required for each system. Details may include weight of flashing, solder and flux, lead, and other materials required for piping installation
- Fitting quantity sheet – documents the number and size of each fitting
- Valve and device quantity sheet – documents the amount, size, and type of each valve and device

Temperature and pressure relief valve nameplate

A temperature and pressure relief valve nameplate should display the following data:
- Date of manufacture built into the serial number. The date of manufacture helps to determine date of service
- American Gas Association (AGA) or American Society of Mechanical Engineers (ASME) approval stamp
- The temperature setting of the valve, for example 210°F
- The pressure setting of the valve, for example 125 psi. The set pressure of the pressure relief valve must not exceed the maximum allowable working pressure (MAWP) marked on the water heater
- The relief capacity of the valve. The minimum relieving capacity of the safety relief valve must not be less than the maximum input

Takeoff sequence

The takeoff sequence is the order of taking quantities from the drawings and specifications. The following is the most effective takeoff sequence for basic plumbing projects in residential and commercial buildings:
- Plumbing fixtures (lavatories, water closets, showers, sinks)
- Plumbing equipment (water heaters, boilers)
- Sanitary waste and vent system (piping, fitting, devices) below grade
- Sanitary waste and vent system (piping, fitting, devices) above grade
- Storm system (piping, fitting, devices) below grade
- Storm system (piping, fitting, devices) above grade
- Hot and cold water system (piping, fitting, devices, valves)
- Natural gas system (piping, fitting, devices, valves)
- Fire standpipe system (piping, fitting, devices, valves)
- Site work

Backflow prevention device nameplate

A backflow prevention device nameplate should display the following data:
- Date of manufacture and a serial number
- The manufacturer's name
- ASSE approval stamp
- Maximum working pressure, for example 150 psi (1034 Kpa)
- Temperature range, for example 32° F – 140° F (0° C to 60° C)
- Hydrostatic test pressure, for example 300 psi (2069 Kpa)
- End detail size, for example ¾ inches, threaded ANSI B2.1
- Fluid type, for example water
- Material of main valve body, for example bronze
- Elastomer type, for example nitrile
- Weight of the device, for example 5 lbs

Water heater nameplate

A water heater nameplate should display the following data:
- Date of manufacture and a serial number
- The manufacturer's name
- The model designation and electrical characteristics, for example 2000 watt (120volt)
- American Gas Association (AGA) or American Society of Mechanical Engineers (ASME) approval stamp
- The heat input, for example, 199,900 Btu/hr
- The capacity, for example, 119.9 gallons
- The temperature setting, for example 210°F
- Continuous hot water rate, for example 7.4 gallon per minute
- Assembled measurements and weight, for example Assembled Depth – 11.125 inches, Assembled Height – 25.375 inches, Assembled Weight – 55.0 lbs, Assembled Width – 14.875 inches
- Rough in data, for example ¾ inches NPT at 5¾ inches c-c (cold inlet to hot outlet)

Lift station nameplate

A lift station nameplate should display the following data:
- Date of manufacture and a serial number
- The manufacturer's name
- System type, for example Simplex or Duplex
- Capacity or flow, for example 30,000 GPM
- Head, for example 250 feet
- Minimum efficiency, for example 70 percent

- Minimum motor horsepower required, for example 2hp
- Electrical characteristics, for example 230 Volts, 230 Hertz, 12.0 amps
- Maximum pump operating speed
- Minimum suction size
- Minimum discharge size, for example 24 inches
- Minimum solid size required to pass through impeller
- Liquid to be pumped, for example abrasive fluid
- Pumping temperature, for example 190°F
- Specific gravity at pumping temperature

Isometric drawings

Isometric drawings are a method of visual representation of three-dimensional objects in a two-dimensional plane.

Isometric drawings are positioned along the three axes, x, y, and z, and are scaled. In plumbing, isometric drawings serve as a means of communication between plumbing professionals as they help to visualize data which otherwise might be difficult to comprehend. Isometric drawings consist of lines and symbols that represent piping and fittings. Abbreviations used in isometric drawings refer to specific fixtures. For example, LAV stands for lavatory, WC for water closet, VTR for vent through the roof. Isometric drawings are used to estimate the cost of a job as well as a reference during the job.

Dimensioning

Dimensioning is a term that refers to placing measurements on the isometric drawing. Dimensions should be taken from the centerline of a pipe of a fitting to the centerline of another pipe or fitting. They can also be taken from a pipe to a part of the building. First, it is necessary to make a sketch with all the dimensions recorded on it. Only the most essential dimensions should be included in the final isometric drawing. The measurement expressed in inches should be placed between two opposite arrows to indicate its relation to the points at the end of the arrows. Extension lines should be provided to clarify the point from which the measure is to be taken. The extension lines should not touch the object represented in the isometric drawing.

Mensuration

Mensuration is measurement of geometric figures including length, angle measure, area, and volume. Taking adequate and precise measures ensures the high quality of the work and efficiency of the installed system.

The following are the steps for mensuration:
- Measure the space where the system is going to be installed. For example, the area of the floors, the height and width of the walls, the pitch of the roof should be measured for the domestic plumbing system setting
- Make calculations to determine the required pipe lengths
- Take measures again before cutting the pipe, installing pipe, fittings, fixtures, and appliances to ensure that there are no errors further in the project

End-to-end measurements

End-to-end measurements are the actual lengths of the pipe to be cut. The following are the steps to obtain end-to-end measurements:
- Measure the space where the work is going to be performed or the object the work is going to be performed on. Pay special attention to the pipe connections and outlets
- After taking each measure, record measurements in a notebook
- Make calculations to arrive at center-to-center, end-to-end, and end-to-center measurements. These measurements include measurements along the pipe and the fittings
- Reduce the end-to-center and center-to-center measurements to allow for the space that the fittings will take up

Measuring tools

The following are the most commonly used measuring tools:
- Six-foot folding rule – line up the end of the rule with the end of the object, then read the length off to the closest division on the rule
- Steel tape – hook up to the end of a board or a pipe, read off at the other end
- Level – place upon or against the surface of the object; observe a bubble in a liquid-filled glass chamber
- Framing square – used for making square cuts on pipe and lumber, setting flanges and welded fittings
- Plumb bob – fasten one end to a solid object and allow the weight to hang; find the vertical reference point or the spot directly below the fastening point
- Soapstone – used to make marks on dark surfaces

Codes and regulations precedence

The uniform rule of code and regulation precedence is that the more specific regulations have precedence over more general, and the more restrictive rule over the less restrictive. For example, the local code for trenching prevails against the International and Uniform Code. Similarly, the manufacturer's instructions take precedence over the general regulations on the installation of various plumbing fixtures.

The following is the order of precedence of plumbing codes and other regulations arranged from the more general to the more specific:

- The International and Uniform Plumbing Code
- ADA and ANSI regulations
- Codes regulating various equipment and systems, such as boilers and pressure vessels, and gas systems
- The State Plumbing Code
- The local building and plumbing codes
- Specific project ordinances
- Manufacturer's instructions

Bar chart schedule

Scheduling is the ability to get the subcontractors on the project at the correct time and in the correct sequence. Most contractors use Gantt (bar) chart schedules. They provide a graphical overview of the project and each task, its duration, relationship between other tasks, and the start and finish dates.

Contractors start with a base schedule and then make adjustments during the process of working on a project. Various computer software programs allow easy creation of schedules and management of projects.

The contract schedule should be formatted as a bar chart showing continuous flow from left to right. Specific calendar dates shall be clearly and legibly shown for the start and finish of each work activity.

Pipe Cutting and Joining

Concrete pipe cutting

Concrete pipe is brittle and has a hard outer skin with a softer center core, therefore care should be taken while cutting concrete pipe by hand. To cut a concrete pipe with a hammer and cold chisel, follow this procedure:
- Tap the pipe with a brick hammer to check if the pipe is sound. A sound pipe gives a clear ringing note, an unsound pipe a dull note
- Mark the pipe with chalk at the length required
- Use a brick hammer or a hammer and cold chisel to chip off the outer hard skin along the chalk line
- Stand the pipe on end. Fill the pipe with sand or well-tamped topsoil
- Continue to chip away the softer centre core along the groove, working round and round the pipe until it breaks along the groove

Cast-iron pipe cutting tools

Tools used to cut cast-iron pipes include:
- Squeeze type cutter – used to cut thin-walled no-hub soil pipe; does not crush pipe like the ratchet cutter
- Ratchet cutter – used to cut extra heavy weight soil pipe no more than 6 inches in size; allows to make a cut with a minimum effort in confined spaces
- Hydraulic cutter – used to cut large sizes of soil pipe
- Caulking hammer – used in conjunction with a cold chisel to make cuts
- Cold chisel – available in a variety of configurations; when struck with a caulking hammer, they make short work of cutting and forming metal, including cast-iron, bricks, tile, concrete, cinder blocks, and stone

The squeeze type, ratchet, and hydraulic cutters are equipped with chains that have a cutter wheel at each link. The chain is wrapped around the pipe, drawn tight by a ratchet or hydraulic action, the handles are pushed together, and the pipe breaks.

Pipe cutters use

Pipe cutters are equipped with one to four wheels that serve as cutting devices. The more wheels the cutter has the less room is needed to cut the pipe. Most pipe cutters are equipped with only one wheel that provides more room for rollers to keep the pipe aligned. The disadvantage is that the cutter has to be passed all the way around the pipe to cut it through.

The following are the steps for cutting the pipe with the pipe cutter:
- Place the pipe in the center of the wheels and rollers
- Rotate the T-handle until it comes in contact with the pipe
- Rotate the cutter around the pipe while turning the T-handle ¼ turn for each revolution around the pipe
- Do not cut through the pipe threads with the cutter to prevent damage to the tool

Hacksaw use

Standard hacksaws come with blades of different sizes, for example, 8 inches, 10 inches, and 12 inches. Blades are made of high quality steel and have between 18 and 24 teeth to the inch. The following are the rules for cutting a pipe with a hacksaw:
- Firmly hold the saw and start cutting with one hand using the thumb on the other hand as the guide
- Hold the thumb high and take light strokes to avoid cutting yourself with the blade
- Once the cut is started, hold the frame lightly with the other hand
- Do not bear down on the saw when drawing the blade back

Thread cutting

The following elements are important for thread cutting:
- The lip – the cutting edge of the tool slanted at a specific angle that differs for each metal. For example, the angle for steel pipe is 15-20°, the angle for brass, copper, and wrought iron is 25°
- The lead – the angle that is machined or ground on the first three threads of each die. The lead enables the die to start on the pipe and distributes the cutting equally on all threads
- Chip space – the space in the die holder in front of the die; allows the chips to curl. Chip space should be large enough to prevent chips from packing and tearing the threads
- Clearance – the distance between threads on the die and on the pipe
- Cutting oil – lubricates the die and reduces the heat caused by friction

Pipe reamers

When the pipe is cut, metal burrs appear on the surface. Pipe reamers remove these burrs. The most common reamer is of the ratchet type with capacity to ream pipe from ¼ inch to 4 inches. The following are the steps for reaming the pipe with the pipe reamer:

- Use the palm of one hand placed against the back of the reamer to insert the reamer in the pipe opening
- Use the other hand to rotate the reamer clockwise
- Do not remove more metal than necessary to obtain the full inside diameter of the pipe

Pipe threaders

Pipe threaders can be of two types, hand threaders and power threaders. Usually, fixed-die-type threaders or ratchet-type threaders are used to thread pipes that do not exceed the diameter of 2 inches. Power threaders are used to thread large quantities of pipe.

Specialized tubing cutters

The specialized tubing cutters can be used to cut plastic pipe and tubing in confined spaces. The specialized tubing cutters are divided into three groups:

- Large internal cutters – used to cut off the excess length of waste tubing stubbed through the floor for a water closet or a shower below the level of the closet flange or the shower strainer. Large internal cutters can cut pipe from 2 to 4 inches in diameter
- Small internal cutters – used to trim off water supply pipes; is able to cut closer to the wall as any other cutter; can cut pipe from ½ to ¾ inch in diameter
- Midget tubing cutters – used in extremely closed spaces; can cut tubing up to 15/16-inch outside diameter

Jab saw can also be used in places where a regular hacksaw cannot be used.

Plastic pipe cutting tools

The following tools are used to cut and ream plastic pipe:

- A blade cutter – the cutting blade forces its way into the wall of the pipe until it is cut through; used for cutting large diameter pipe
- A handsaw and a wooden miter box – used to hold the pipe in place while the pipe is cut with a hand saw
- A hacksaw – makes a square cut but leaves a ragged burr

- A vise – equipped with plastic-covered jaws to protect the pipe
- A reaming tool – used to remove burrs on the inside and outside of the pipe; available in sizes for pipe from ½- to 4-inch in diameter

Cutting and reaming steel pipe

Power tools
The following are the power tools used to cut and ream steel pipe:
- A blade type geared cutter – removes the metal as it rotates around the outside diameter of the pipe; used to cut pipe between 2½ and 12 inches in diameter
- Portable pipe threading machine – complete with cutter, reamer, and die head mounted on a tripod stand; used to cut pipe 2 inches in diameter and less
- Large diameter pipe cutting machine – complete with a built-in oiler, cutter, reamer, and adjustable die head; used to cut pipe between 2½ and 4 inches in diameter
- Abrasive saw – used to cut large diameter pipe and other kinds of metal on the job site

Hand and power cutters
The following are the hand tools used to cut and ream steel pipe:
- A pipe vise – used to hold the pipe firmly in place
- A wheel pipe cutter – used to make a square cut on the pipe; the wheel squeezes the metal and forces it ahead of the cutter until the pipe is cut through its wall thickness
- A hinged four-wheel pipe cutter – rotates a little more than 90° to make a cut; can be used on an installed pipe; cuts pipe up to 12 inches in diameter
- A spiral pipe reamer – used to remove ridges left by the wheel pipe cutter
- Half-round file – used to ream pipe larger than 4 inches in diameter

No-hub cast-iron soil pipe fittings

No-hub cast-iron pipes and fittings enable plumbing specialists to join pipes and fittings in a faster method. No-hub cast-iron pipes and fittings are joined with mechanical gasket joint. The pipes are cast in 10-foot lengths and are manufactured in sizes from 1½ inch to 10-inch inside diameter. The following table illustrates types of no-hub cast-iron soil pipe fittings:

1/16 Bend	1/8 Bend	1/6 Bend	1/4 Bend	Long Sweep	Short Sweep
Sanitary Tee	Wye		Sanitary Tap cross		Short Reducer

Bell and spigot

Bell and spigot cast-iron soil pipe and fittings are equipped with a bell or hub cast at the end where the spigot or plain end of another pipe or fitting is inserted to make a joint. The hub and the pipe are sealed with a caulked lead and oakum joint, or a mechanical compression joint. The pipes are cast in 5-foot and 10-foot lengths for single-hub pipes and 30-inch and 5-foot lengths for double-hub pipes.

<u>Making a bell-and-spigot</u>
Concrete method: The following are the steps for making a cement joint on vitrified clay pipe:
- Grade the bottom of the trench properly
- Hollow out a space under each bell to allow the pipe to rest on its side
- Insert the spigot end of the pipe into the slant
- Force a piece of oakum into the bell of the slant to center the pipe and prevent the cement from being forced inside the pipe and form an obstruction
- Mix one part of cement and two parts clean sand. Add just enough water to make a stiff mortar
- Force cement into the bell and trowel until smooth
- Fill the space around the joint and the pipe with soil to prevent the pipe from moving when the next joint is made
- Clean the inside of the pipe after each joint is made

Bituminous method: The following are the steps for the bituminous method for making a bell-and-spigot joint on vitrified clay pipe:
- Grade the bottom of the trench properly
- Hollow out a space under each bell to allow the pipe to rest on its side
- Insert the spigot end of the pipe into the slant
- Secure a joint runner and coat it with soft mud to prevent the joint compound from sticking
- Place the runner around the joint
- Place mud around the top packing it and making space for the heated compound
- Pour the heated compound in the space provided and let it set
- Remove the runner; pack the dirt around the pipe

Caulked soil pipe joint

The following are the steps for making a caulked soil pipe joint:
- Cut the soil pipe to the required length
- Clean the hub and spigots of all foreign materials, moisture, grease, etc
- Assemble the pipe making sure to align and space the joint properly
- Yarn and pack oakum into the hub approximately 1 inch from the top avoiding protruded loose fibers
- Use a ladle to pour the molten lead into the vertical soil pipe. Place an asbestos running rope around the horizontal pipe and clamp tightly at the top to form a gate for pouring lead
- As soon as lead turns solid in both vertical and horizontal pipes, drive the lead down with the caulking hammer and iron to set the joint
- Caulk around the inside edge of the soil pipe joint with the inside caulking iron
- Caulk around the outside edge of the soil pipe joint with the outside caulking iron.

No-hub soil pipe joint

The following are the steps for making a no-hub soil pipe joint:
- Detach the neoprene sleeve from the stainless steel clamp assembly
- Slide the stainless steel clamp assembly onto the pipe
- Place the spigot ends of the pipe or fitting inside the neoprene gasket and push until the ends bump against the pipe or fitting separator ring
- Slide the stainless steel clamp assembly back onto the neoprene gasket
- Tighten the screw clamps with a torque wrench to a minimum of 48 inch-pounds of torque. The screw clamps can be tightened with other tools, for example, a 5/16-inch nut driver, a ¼-inch ratchet with a 5/16-inch socket, or an electric nut driver

Lead can be obtained in 25-pound bars. The lead should be carefully added to the molten pot as damp or frozen lead will explode if added to a lead pot containing partially molten lead. Always preheat damp or frozen lead at the side of the furnace before adding it to the pot with molten lead. The pre-heated lead is then heated on the furnace to the degree when it melts but is not red-hot. Overheating lead should be avoided because the metal burns into a slag. Then, the lead is poured into a vertical soil pipe joint with the ladle. The molten lead should not stick to the ladle.

Compression gasket pipe joint

The following are the steps for making a compression gasket pipe joint:
- Clean the hub and spigot of all foreign materials, such as dirt, mud, gravel, moisture, grease, etc.
- Remove the sharp edge of the cut pipe to avoid jamming the pipe against the gasket. The sharp edge can be removed by peening with a hammer or rasping with a file
- Insert the gasket into the hub with one of the following methods: 1) folding; or 2) bumping
- Lubricate the gasket with pipe lubricant
- Push or pull the spigot through both seals of the gasket
- Force the spigot partially out through the first seal to install fittings. Drive it back with a lead maul

Bumping gasket method

The bumping gasket method is suitable for installing gaskets of larger diameter as it is difficult to fold large rubber compression gaskets. To insert the gasket into the hub using the bumping method, do the following:
- Place the rubber compression gasket in the soil pipe hub. Make sure the ring of the gasket is secured in the hub groove
- Use the heel of your hand or a wooden board to bump the gasket
- Strike the fitting against a board or the floor to make sure that the rubber compression gasket is inserted tightly into the fitting and cannot be bumped out

Folding gasket method

The folded gasket method is suitable for installing gaskets of smaller diameters as they are the most difficult to insert. To insert the gasket into the hub using the folding method, do the following:
- Hold the rubber compression gasket upright at the bottom with your thumbs. Use your thumbs to fold the bottom of the gasket through the top as if you were turning it inside out
- Put the folded rubber compression gasket in the soil pipe hub. Make sure the ring of the gasket is secured in the hub groove
- Release the gasket. As you release the gasket, it will unfold into the hub of the soil pipe

External pipe thread

The following are the steps for making an external pipe thread:
- Place and secure the galvanized iron pipe in the vise or pipe machine jaws
- Cut the pipe to the required length. Be aware that the inside diameter can be reduced by as much as ¼ of an inch
- Ream the inside of the pipe to the original diameter
- Choose the proper thread length to fir the pipe
- Use the die to thread the pipe to the required length. Lubricate the die as you are threading the pipe
- Remove the die and clean the thread
- Apply suitable pipe joint compound to the external thread
- Place the fitting onto the pipe thread and tighten the joint with a pipe wrench

Galvanized steel threaded joints

Galvanized steel pipes, malleable back iron pipes, brass and copper pipes, cast iron pipes, and thick-walled plastic pipes are joined to the water supply, drain, or vent fittings with a threaded joint.

The commonly used thread used for this type of joint is the American Standard Taper Pipe Thread, abbreviated as NPT. The NPT is tapered ¾ inch per foot of thread length to form a tight leak-proof joint between the pipes and the fittings. The NPT characteristics are classified in the American Standard Taper Pipe Thread Table. There is no need to make the internal, or female, pipe thread, as the pipes are usually manufactured with these threads. Only the external, or male, pipe thread should be made.

Paste-type solder rules

The following rules should be observed while working with a paste-type solder:
- In addition to the paste, apply wire solder in order to fill voids and displace the flux. If the wire solder is not applied, a continuous bond may not develop
- Mix the solder paste mixture thoroughly, especially if it has been kept in the can for a while, as the solder settles quickly to the bottom
- Clean the tube manually as the flux alone cannot clean the tube properly
- Apply the flux to the clean surface
- Make sure to remove any excess flux whose main functions are to remove residual oxides, promote wetting, and to protect the surface from oxidation when heated

Solder selection

There are different types of solder, so the selection depends on the working pressure and temperature of the plumbing system. Some of the most commonly used types of solder are 50-50 and 95-5 wire solder. The 50-50 wire solder is made of 50 percent of tin and 50 percent of lead and is suitable for moderate pressures and temperatures. The 95-5 wire solder is made of 95 percent of tin and 5 percent of antimony and is used for higher pressures and temperatures, or areas where greater joint strength is required. The 95-5 wire solder melts at higher temperature than the 50-50 wire solder, so it is more difficult to use. Its melting temperature also affects the pasty range of the solder that is the temperature range between which solder is neither liquid nor solid. Paste-type solder can also be used in addition to wire solder.

Impact flared joint steps

The following are the steps for impact flared joint:
- Cut the tube to the required length
- Remove burrs
- Place the coupling nut over the tube end
- Insert the flaring tool into the tube end
- Drive the flaring tool with a hammer and expand the end of the tube to the required flare
- Place the fitting squarely against the flare to assemble the joint

Screw-type flared joint steps

The following are the steps for screw-type flared joint:
- Follow the previous steps 1 to 3
- Clamp the tube in the flaring block to place the tube end above the face of the block
- Put the yoke of the flaring tool on the block to place the beveled end of the compression cone over the tube end
- Form the flare by firmly turning the compressor screw down
- Remove the flaring tool and assemble the joint

Solder joint steps

The following are the steps for making a solder joint:
- Cut the tube end square to the required length with a tubing cutter or a hacksaw
- Ream the cut end of the tube
- Use sand cloth to clean the tube
- Use a fitting brush, steel wool, or sand cloth to clean the fitting socket
- Apply flux to the tube end and the fitting socket

- 35 -

- Put the pipe and the fitting together
- Apply heat to the tubing for a short time. Then, apply heat to the fitting melting the solder placed at the joint of the tube end fitting
- Stop heating and feed solder into the joint until a ring of solder appears at the end of the fitting
- Use cloth to remove excess solder while it is still pasty
- Leave a fillet around the end of the fitting

Solvent weld joint

Solvent weld joint is made by producing a welded system with solvent bonding. The primer and solvent are used to soften the material on the outside of the pipe and inside of the fitting. If the solvent weld joint is made properly, the pipe and the fitting fuse turning into a tightly bonded piece that is as strong as the pipe. It is important to follow these rules while making a solvent weld joint:
- Choose the right primer and solvent. Both materials should be formulated specifically for the type of plastic to be used in order to produce solid bonding. The importance of primer should not be underestimated; primer prepares the pipe and fitting for solvent application by cleaning and softening up the surfaces
- Check for an interference fit. The pipe should fit approximately halfway into the fitting socket
- Carefully prepare and install the solvent weld joint

Mechanical compression joint parts

The following are the parts used for mechanical compression joints:
- A compression joint fitting
- A compression ring
- A compression nut
- The plain end of the copper tube

Copper compression joint

The following are the steps for making a copper compression joint:
- Cut the tube to the required length
- Use sand cloth to clean the tube of all foreign material
- Slide the compression nut onto the copper tubing
- Slide the compression ring onto the copper tubing
- Slide the mechanical joint fitting over the tube end
- Compress the compression ring into the tubing and seal the joint by tightening the compression nut onto the fitting with a wrench

Installing an insert fitting joint

To install an insert fitting joint for plastic tubing, it is necessary to push it into the ends of the pipe and secure with stainless steel clamps. The following are the steps that are required for insert fitting joint installation:
- Cut the tubing to the required length with a plastic tubing cutter, knife, or saw
- Join the tubing by slipping the loose stainless steel clamps onto the tubing ends and pushing the insert fitting into the tube ends until it bumps into the fitting shoulder
- Secure the fitting by sliding the stainless steel clamps over the serrated section of the fitting and tightening the clamps with a screwdriver

Joining PVC pipes

The following are the steps for of joining PVC pipes with the solvent method:
- Cut the pipe squarely with a miter box or a sharp tube cutter
- Clean and smooth the end of the pipe removing all burrs
- Check for interference fit by fitting the pipe halfway on a dry joint
- Apply suitable primer to clean and dry surfaces first to the inside, then to the outside of the pipe
- Wait 5-15 seconds
- Apply solvent cement on the surfaces still wet from the primer. First apply the cement to the inside, then to the outside of the pipe. Make sure to apply a full, even coat of the solvent cement
- Fit and position the pipe and its fitting. Assemble the pipe and the fitting immediately to seal the joint
- Check for the correct bead
- Clean off excess cement

International and State Code requirements

The following are the International and State Code requirements for joining plastic tubing:
- The joints for plastic piping and fittings should be made in one of the following methods: a) solvent-cement method; b) heat-fusion method; or c) with compression couplings and flanges
- Threading plastic pipe is prohibited
- Heat-fusion or mechanical joints should be made on polyethylene tubing
- Heat-fusion joints should be made on polyolefin piping or tubing
- Mechanical fittings used with polyethylene piping must comply with ASTM D 2513 category 1, full-restraint, full-seal joints
- Plastic gas piping should be joined with mechanical compression fittings that must be approved for use with the pipe material and gas carried

- An internal tubular rigid stiffener must be used in conjunction with the compression-type mechanical fitting in gas pipes
- Split tubular stiffeners must not be used

Flare fitting joint

The following are the steps for installing a flare fitting joint for flexible plastic tubing:
- Cut the tubing to the required length with a plastic tubing cutter, knife, or saw.
- Slide the flare nut on the tubing
- Insert the pilot plug of the flaring tool into the fitting
- To complete the flare, rotate the tool 5 to 10 revolutions with the clamping pliers
- To assemble the joint, place the flared end of the tubing against the fitting and tighten the flare nut onto the fitting. To do this, two wrenches should be used: one wrench for the fitting and one for the flare nut

Ductile iron pipe

Ductile iron pipe is manufactured from a steel-iron alloy with flanged, mechanical, or push-on joints. The ductile iron pipe comes in 18-inch lengths and is classified into classes according to the pressure or wall thickness.

Pressure classes are 50, 100, 150, 200, 250, 300, and 350 psi; they indicate the pressure under which a specific class of pipe is designed to operate. Ductile iron pipe comes with the following wall thickness: 50, 51, 52, 53, 54, and 56.

Ductile iron pipe that comes in contact with soft water should be insulated with a cement lining to prevent tuberculation. Ductile iron pipe is highly resistant to corrosion and physical damage from backfill operations and other stress.

Unloading and handling plastic pipe

The following are requirements for unloading and handling plastic pipe:
- The site for unloading plastic pipe should be flat and clean, free of nails, rocks, and other sharp debris that might damage the pipe
- If it is impossible to provide a clean site for unloading and handling plastic pipe, wooden supports should be provided at no more than 3-foot intervals
- If mechanical handling is used, metal slings and hooks should not be used. Instead, standard pipe rope should be used
- When stored in storage racks, plastic pipe should be supported on at least six evenly distributed supports for each length of pipe

Steel pipe threaded joint

Threading is used to join thin wall steel pipe. Threading is done manually or with power-operated threading devices equipped with hard tempered steel dies which come in various sizes to produce tapered pipe threads. Threads should be gauged according to the standards established by the American Standards Association. To join threaded pipe, follow these steps:

- Clean the male threads of the pipe and the female threads of the fitting with a hard wire brush
- Apply the joint compound to the male end
- Manually align and tighten the pipe and the fitting
- Complete the joint by tightening the pipe and the fitting with a pipe wrench. Use Teflon tape if required

Ductile iron pipe joints

The following joints are made on ductile iron pipe:

- Flanged joint – made by threading flanges onto the threaded pipe ends. To make a flanged joint, place a rubber or special application face gasket between the two flanges; bolt the flanges together
- Push-on joint – a compression-type joint. To make a push-on joint, insert a rubber or neoprene gasket into the hub or bell end of the pipe; use a special joining tool to slip the spigot or plain end of the pipe into the bell end
- Mechanical joint – a cross between a flanged joint and a push-on joint. To make a mechanical joint, insert the spigot end of the pipe into the bell end with a rubber ring gasket and an outer ductile iron retainer gland; bolt together the bell end and the spigot end of the pipe

Flue pipe installation

The following are the guidelines for twin-wall flue pipe installation:

- Use 5-inch twin-wall metal flue pipe (not single or flexible pipe) to connect the concrete gas flue block system to a gas flue terminal outlet
- Maintain a minimum of 1 inch away from combustible materials (measured from the inner flue pipe)
- Do not cut twin-wall flue pipe

Joining
The following are the guidelines for flue pipe joining:

- Join the pipe sections by aligning the dimples on the couplings
- Push fitting together with the male coupling pointing up towards the terminal
- Twist the upper pipe clockwise 1/6 turn to complete the joint

Use angled flashing kit when going through the roof space to provide weatherproofing. Fit a gas vent terminal when terminating through the roof space. Twist-lock the terminal to the final section of pipe with the standard twist-lock connection.

Steel pipe grooved joint

Grooved mechanical joint is a low cost method that eliminates time-consuming installation requirements used in other joint methods. Grooved joints are used in fire protection piping, building service systems, potable, heating, cooling, and condenser water systems. The fittings are manufactured in black or galvanized steel, for use with plain end, cut grooved, roll grooved, or bevel end steel pipe. Grooved pipe and fittings are manufactured in configurations that allow them to form a complete system only if specific matching pipe and fittings are used. Special tools are required to cut or roll groves, and to drill outlets in the pipe wall where hook-type outlets are substituted for tees or welded nozzles.

Polyethylene pipe

Polyethylene pipe is a flexible plastic pipe used for cold water underground plumbing installations, for example, wells, sprinklers, and water supply. It is usually set outside the foundation walls of the building. Polyethylene pipe should be protected from freezing; that is why it should be laid in a trench that is at least 12 inches below the frost line. Since polyethylene pipe has a high expansion rate it should be allowed to snake or wind down the trench. After the pipe is laid, the trench should be carefully backfilled to a depth of 6 inches above the pipe. A pressure test should be conducted before backfilling further. Polyethylene pipe is available in coils of 500 feet and can sustain pressures between 80 and 160 psi.

Insert filing joint

The following are the steps to insert filing joint on polyethylene pipe:
- Use a saw or a tubing cutter to cut the pipe squarely
- Use a half-round file to ream the end of the pipe
- Place two hose clamps onto the end of the pipe
- Press the pipe onto the fitting without any pipe dope. If the fit is too tight, put the end of the pipe in a weak soap solution or into very hot water
- Tighten the hose clamps placing the screws 180° apart so that the tension is distributed evenly

Polyethylene pipe can also be cold-flared with a flaring tool. The flared joint is assembled in the same way as any other flared fitting.

Asbestos pipe joints

Asbestos pipes with similar diameter can be joined with sleeve couplings of the same material as the pipe and seated with approved gaskets or rings. Asbestos cement pipe and metal pipe should be connected with a proper adaptor coupling that should be joined in a caulking method. The joints should be caulked, firmly packed with oakum, and secured with pure lead to a minimum depth of 1 inch. The lead should be poured in one move, and the joint should be caulked tight.

Clay pipe joints

Lengths of clay pipe in building sewers are joined on both bell and spigot ends. The joints should be made with resilient materials that conform to the standards established by Type III of ASTM.

Glass pipe

Glass pipes and fittings are similar to metal pipe in durability, however, they should be handled with care and installed without strain. Glass pipe and fittings should be stored in boxes until ready to be used. Any scratches or nicks on the surface of the pipe may cause breaks and leaks later. Glass pipe can be installed inside and outside. If installed underground, it should it insulated with materials recommended by the manufacturer. If glass pipe is exposed, it should be protected with wire mesh, angle irons, or channels. Glass pipe should also be protected from pressure surges by installing pressure relief valves, or air chambers. The expansion rate of glass pipe is ¼ inch per 100 feet of pipe per 100° temperature difference.

Glass pipe joints and fittings
The most common method for glass joints is beaded-end pipe and fittings. The following are the steps to make a glass pipe joint:
- Use a metal compression coupling with a corrosion-resistant Teflon gasket to receive one beaded end and one plain cut end
- Tighten the coupling with two bolts which are part of the metal housing of the coupling

The joint can be also made with couplings designed to receive two beaded ends. A number of fittings are available for making glass pipe joints. For example, the following fittings can be used: elbows, bends, tees, wyes, reducers, vent stack increasers, P-traps, cleanouts, etc. Conical joint glass piping is coupled with flanged, gasketed joints.

Rough-in plumbing

Installation of the rough-in plumbing consists of pre-installation of all the plumbing parts used for fixtures to be installed later and includes installation of drainage, vent, water supply piping, and necessary fixture supports. The rough-in plumbing is usually installed after all the framing of the house is completed and the roof is covered. Before beginning rough-in plumbing installation, it is necessary to locate the various plumbing fixtures and appliances. This can be done by studying carefully appropriate sheets of the house plans. Also, the specifications should be checked to determine the type, size, shape, and material of the pipes, joints, cleanouts, traps, and other parts of piping to be installed.

Perform plumbing systems installation

Potable water supply system

Parts of the potable water supply system include:
- Potable water – water purified according to the requirements of the state board of health
- Potable water supply system – water service pipe, water distribution pipes, connections, fitting, control valves, and trimming inside or outside the building
- Water main – conveys potable water from the municipal water supply source for the public or community use
- Corporation stop – a valve on the water main pipe to which the building service water is connected
- Water service – leads from the water supply to the water distribution system of the building
- Curb stop – a valve on the water service placed near the curb line
- Water meter – a device measuring the volume of water used
- Water distribution pipe – conveys water from water service to fixtures and appliances
- Fixture supply – connects water supply pipe to fixture branch
- Fixture branch – connects fixture supply to water distribution

Fixture unit

A fixture unit is a unit used to measure the rate of water flow, equal to one cubic foot of water (roughly 7.48 gallons) per minute. A fixture unit is the basis for plumbing design of both water supply and waste water systems. Each fixture is assigned its own particular value; the distinction is made for similar units with different waste pipe sizes. Fixtures that flow continuously, for example, air conditioners and sewage ejectors, require addition of 2 more units for each gallon per minute of flow. The maximum fixture unit load is the greatest number of fixture units that may be connected to a specific size of a house drain, horizontal branch, vertical soil, or waste stack.

Water supply pipe minimum sizes

Selection of water supply piping is restricted to the following minimum sizes:
- ¾-inch size pipe – for any building from the street to the water meter
- ¾-inch size pipe – for the first section of water supply piping within the building
- ¾-inch size pipe – to a sill cock or lawn faucet
- ¾-inch size pipe – from cold water supply to a water heater

- ¾-inch size pipe – for the first section of hot water piping on the outlet side of a water heater
- no less than ½-inch size pipe – for concealed piping
- ½-inch size pipe – for no more than three fixtures in one bathroom or house
- ½-inch size pipe – for fixture branch piping
- ⅜ or ¼-inch size pipe – for individual fixture water supply piping that is not concealed and does not exceed 30 inches in length

Factors

In order for the plumbing fixtures and appliances to operate properly the building water supply system must provide sufficient potable water. The following five factors determine the sizing of water supply piping:
- Available pressure – usually between 45-60 psi; should not exceed 80 psi
- Demand – also called flow rate; the volume of water in gallons each fixture and appliance uses per minute
- Length of piping – causes the pressure loss due to friction
- Height of the building – causes the pressure loss due to the height the water must flow
- Flow pressure required at the top floor – minimum flow pressure required for a fixture or appliance to function properly; varies from 8 to 25 psi

Fixture water supply pipe

Selection of fixture water supply piping is restricted to the following minimum sizes:
- Bathtubs – ½ -inch size pipe
- Bidets – ⅜-inch size pipe
- Combination sink and tray – ½ -inch size pipe
- Dishwasher – ½ -inch size pipe
- Drinking fountain – ⅜-inch size pipe
- Kitchen sink – ½ -inch size pipe
- Laundry – ½-inch size pipe
- Lavatory – ⅜-inch size pipe
- Shower – ½-inch size pipe
- Flushing rim sink – ¾-inch size pipe
- Service sink – ½-inch size pipe
- Water closet with flush tank – ⅜-inch size pipe
- Water closet with flush valve – ½-inch size pipe
- One piece water closet – ½-inch size pipe
- Urinal with flush tank – ½-inch size pipe
- Urinal with flush valve – ¾-inch size pipe
- Hose bibs – ½-inch size pipe
- Wall hydrant – ½-inch size pipe

- 44 -

<u>Sizing</u>

The following are the steps for sizing water supply pipe in large installations using appropriate tables in the local plumbing code:

- Total the cold water supply fixture unit (cwsfu) demand for the entire building assigning values from 10 to 40 for each flushometer valve starting with the most remote
- Total the hot water supply fixture unit (hwsfu) demand
- Determine the longest developed piping length
- Determine the difference in elevation between source of supply and the highest outlet
- Deduct this difference from the available water pressure. Find this figure in the table
- In the table, find the maximum allowable length matching the figure from Step 3
- Follow to find the water service and water meter size matching the figure calculated in Step 1
- Size the building supply pipe
- Size each piece of pipe from the most remote fixture to the water meter
- Size the cold water supply to the heater and the hot water supply out of the heater
- Add the hwsfu to the cold water main
- Size the hot water supply pipe

Water velocity method

The water velocity method can be used for sizing pipe in buildings three stories high or less. The pressure of water available should be at least 40 psi. The following are the steps to size piping with the friction loss method:

- Obtain the available pressure at the main and the corrosive qualities of water from the local administrative authorities
- Make an isometric drawing of the whole system indicating branches, fixtures, and quick-closing valves
- Identify the total water supply fixture units for each fixture and the hot-and-cold water valves
- Use the minimum pipe sizes from 112b to mark pipe sizes for each fixture starting from the fixture outlet
- Use velocity limitation tables (available in local and standard plumbing codes) to determine the pipe size from the fixture to the service pipe

Friction loss

Friction loss is the rubbing action of the moving water against the interior sides of the pipe. The length of the pipe increases the friction loss proportionally. For example, if the pipe size is doubled, the friction loss is doubled as well. If the interior of the pipe corrodes and becomes rough, the friction loss increases significantly.

Water velocity

Water velocity is a measure of how fast water flows through the piping. Friction of water is proportional to the square of the velocity. Average water velocity equals 8 feet per second under normal conditions. Increased water velocity causes excessive wear, corrosion, and water hammer – a knocking sound in water pipes caused by a sudden change in pressure after a faucet or water valve shuts off.

Both friction loss and water velocity should be considered when sizing water supply pipes.

Vent piping system parts

The following are the parts of the vent piping system:
- Individual vent – a pipe that vents individual fixture traps; may terminate into a branch vent, a vent stack, a stack vent, or the atmosphere
- Branch vent – connects two or more individual vents with a vent stack or a stack vent
- Stack vent – an extension of a soil or waste stack above the highest horizontal drain connected to the stack
- Vent stack – a vertical pipe that provides air circulation to and from the drainage system
- Roof jacket or flange – a protective device preventing rainwater from entering the building around the vent pipe; is installed on the roof terminals of stack vents and vent stacks

Sanitary drainage and vent system parts

The following are the parts of the sanitary drainage and vent piping system:
- Sanitary drainage pipe – removes wastewater from plumbing fixtures and conveys it to the sanitary sewer
- Vent pipe – ventilates a drainage system
- Cleanout – gives access to the pipe for cleaning
- Waste pipe – conveys only liquid waste
- Soil pipe – conveys solid waste
- Stack – any vertical pipe extension
- Drain branch – receives discharge from the waste, soil, and other drainage pipes and conveys it to the sewer
- Fixture drain – the drain from the trap of the fixture to the connection of the drain with another drainpipe
- Fixture trap – provides a liquid seal to prevent emission of sewer gases

Horizontal branch drain sizing

The State Plumbing Codes provide tables for calculating the horizontal branch drain sizing based on drainage fixture unit value and slope. The following example illustrates the sizing of a horizontal branch drain for a building with small-sized facilities.

Drainage Fixture Unit Summary:
- Water Closets – 2, dfu (drainage fixture unit value) – 6
- Lavatories – 2, dfu – 1
- Bathtubs – 2, dfu – 2
- Kitchen sinks – 2, dfu – 2
- Slope – ¼ inch per foot

The number of units should be multiplied by the dfu value; then, the total must be computed. In this case, the total is 22 dfu. This value should be found in the table and checked against the pipe sizes in the left hand column. Considering the slope, the recommended pipe size is 3 inches.

Drainage pipe sizing

The State Plumbing Codes provide tables for calculating the drainage pipe sizing based on the type of fixture, number of fixtures, drainage fixture unit value, and slope. The following example illustrates the sizing of a horizontal drainage pipe for a building with medium-sized facilities.

Drainage fixture unit summary:
- Water Closets – 30, dfu (drainage fixture unit value) – 6
- Lavatories – 28, dfu – 1
- Drinking fountains – 4, dfu – 1
- Urinals – 3, dfu – 4
- 2" floor drains – 4, dfu – 2
- Service sinks – 2, dfu – 3
- Slope – ¼ inch per foot

The number of units should be multiplied by the dfu value; then, the total must be computed. In this case, the total is 238 dfu. This value should be found in the table and checked against the pipe sizes in the left hand column. Considering the slope, the recommended pipe size is 5 inches.

Change in direction

Change in direction is a term that refers to any turns in the length of the pipe line made with various fittings. The selection of the proper fitting depends on the type of the change in direction, and the size of the pipe. The following table illustrates the selection of fittings for specific pipe sizes and types of change in direction:

Change in direction	Pipe size	Fitting
Vertical to horizontal	Less than 3 inches	Long sweep soil ¼ bend or extra long turn 90o drainage ell
Vertical to horizontal	3 inches and larger	Short sweep soil ¼ bend or long turn 90o drainage ell
Horizontal to vertical	All sizes	Soil ¼ bend or long turn 90o drainage ell
Horizontal to vertical	All sizes	Long or short sweep soil ¼ bend or extra long turn 90o drainage ell

Horizontal drainage piping grade

The grade, or pitch, is a slope of a pipe line in relation to a horizontal plane. It is expressed as the fall in a fraction of an inch per foot of the pipe length. Most horizontal drain piping is graded at ¼ of an inch per foot. This grade facilitates the necessary velocity and discharge capacity and prevents the increase or decrease of pressure in the plumbing system. A grade less than ¼ of an inch is used for longer drainage pipes and in cases of insufficient basement of main sewer depth. A grade greater than ¼ of an inch is used to increase velocity and discharge capacity but decrease the depth of waste necessary to instigate the self-scouring action. The following are the minimum grades for specific pipe sizes:

Pipe size	Minimum grade
Less than 3 inches	¼"
3 to 6 inches	⅛"
8 inches and larger	1/16"

Kitchen sink installation

The following are the steps for the installation of a kitchen sink stack in a single-family dwelling:
- Complete installation of the building drain
- Extend a 2-inch pipe above the basement floor to a 1½-inch test tee for a stack base cleanout
- Extend a piece of pipe from the top of the test tee to the top of the concrete block basement wall
- At the top of the wall, use two bends to offset back over the wall to the stud wall in the kitchen
- Extend the pipe to the sanitary tee for the kitchen sink drain
- Extend the top of this tee to a point below the roof and continue through the roof to vent the kitchen sink trap
- At the sanitary tee, extend the pipe over a tapped bend to the sink drain connection

Building drain installation

The following are the steps for the installation of a building drain in a single-family dwelling:
- The building drain is installed after the trench is completed
- A 4-inch soil pipe is equipped with proper joints and pushed out beneath the footing at the front of the building to be later connected to the building sewer
- The building drain is extended to the front main cleanout and to each future fixture and appliance, such as shower drain, kitchen sink drain, water closet and lavatory, laundry tray and floor drains by using the same size piping and joints as the soil pipe
- 4" by 2" wyes are installed first, then pipes
- Each section of the piping should be placed directly in line with the drain openings of each fixture and appliance

Lavatory drain installation

The following are the steps for the installation of a lavatory drain in a single-family dwelling:
- Put a 2" by 1½" no-hub coupling on the right-hand 2-inch opening in the 3" by 2" double wye
- Install a bend in this coupling
- Extend the pipe to the bend and a combination wye with the side opening looking up directly beneath the whole drilled for the lavatory drain
- Cap the bend-wye combination with a blind plug for a cleanout
- Extend the top of the combination fitting to the 1½" sanitary cross that should be centered at 17 inches above the floor
- Attach a tapped bend for each side of the sanitary cross opening

- 49 -

Bathroom installation

The following are the steps for the installation of a bathroom in a single-family dwelling:

- Install a 3-inch main soil stack
- Extend the soil pipe to the drains from the shower, bathtub, and lavatory
- Join a 3-inch sanitary cross to the top of the 3" by 2" double wye
- Cut closet bends to length and join to the side opening of the sanitary cross to provide the water closet waste openings
- Extend a piece of 3-inch pipe from the top of the sanitary cross to a 3" by 2" sanitary cross to be used for the shower and lavatory vent connections
- Extend the stack from the top of the 3" by 2" sanitary cross through the roof

Bathtub drain and vent installation

The following are the steps for the installation of a bathtub drain and vent in a single-family dwelling:

- At the bathtub drain wye, extend the opening to a 1½-inch ⅛ bend. The bend should be centered on the whole drilled for the bathtub vent
- Extend the pipe from the bend to a combination wye and a ⅛ bend lying on its back with the side opening of the combination fitting looking up directly beneath the tub vent whole
- Cap the end of the combination fitting with a 1½-inch no-hub blind plug to provide a cleanout
- Join a tapped tee to the top of the combination fitting for the bathtub waste connection
- Extend the top of the tee to a point below the roof, increase the pipe with a 2" by 1½" no-hub coupling, and continue through the roof

Shower drain installation

The following are the steps for the installation of a shower drain in a single-family dwelling:

- Join the bend at the stack to the opening of the 3" by 2" double wye
- Extend a 2-inch pipe from this bend to the wye for the bathtub drain
- Join the bend and a combination wye with the side opening looking up directly beneath the whole for the shower bath vent
- Cap the end of this combination fitting with a no-hub blind plug for a cleanout
- Set a 2-inch sanitary tee on the top opening of the combination fitting
- Extend the side opening of the tee to a 2-inch P-trap
- Extend the inlet of the trap above the floor

- Reduce the tee with a no-hub coupling. Extend the pipe through the floor to a bend centered at 37 inches above the floor
- Extend the bend to the 3" by 2" sanitary cross and join the cross with a no-hub coupling

Floor and area drains

Floor drains are used to carry away leakage and spilled water in elevator pits, boiler rooms, laundry rooms, and garages. Area drains are used for the same purposes and are installed in driveways and other paved surfaces. Floor and area drains are made of cast iron and are extra heavy to endure the weight of traffic. Plumbing codes in some states require floor drains with flanges. The top of a floor or area drain should be removable for cleaning purposes. Floor and area drains should be equipped with a trap on the waste pipe. The trap should be installed below the frost line to prevent freezing. Floor drains should be caulked to the soil pipe. Area drains should be equipped with a deep seal trap to prevent evaporating. Floor drains should have a handy water supply to replenish the water in the seal in case of excessive evaporation.

Laundry tray waste and vent installation

The following are the steps for the installation of a laundry tray waste and vent in a single-family dwelling:
- Extend the 2-inch sweep ¼ bend installed for the laundry tray above the basement floor line
- Reduce the pipe with a no-hub coupling
- Extend the pipe to a 1½-inch tapped cross which should be centered at 18 inches above the basement floor
- From the top of the taped cross, extend a piece of pipe to a 1½ -inch ¼ bend near the ceiling
- Extend the pipe across the basement ceiling towards the bathroom
- At the center of the sanitary tee, join the pipe to another 1½ -inch ¼ bend
- Join the bend and the tee to complete the laundry tray vent and waste piping

Gas pipe sizing methods

The pipe size of each section of the longest pipe is calculated using the longest length of piping from the initial point of delivery to the most remote outlet and the load of the section.

Branch length method
The pipe size of each section in branch piping is calculated using the longest length of piping from the initial point of delivery to the most remote outlet in each branch and the load of the section.

<u>Hybrid pressure method</u>
The pipe size of each section in higher pressure gas piping is calculated using the longest length of piping from the initial point of delivery to the most remote line pressure regulator. The pipe size from each pressure regulator to each outlet is calculated using the length of piping from the regulator to the most remote outlet served by this regulator.

Closet bends

Closet bends are fittings made of cast iron with a diameter of 4 inches that connect the water closet to the drainage system. Closet bends are joined to the drainpipe with a caulking method. Closet bends come in many different lengths; it is also possible to adjust the length by cutting the bend at the grooves at either end. The bends are equipped with side outlets or tappings on both sides. Although cast-iron closet bends prove to be stronger than bends made of other materials, some local plumbing codes do not allow the use of cast-iron bends claming that they are too rigid and might break the toilet bowl if the building or the stack settles.

Gas pipe sizing

The following are the National Fuel Gas Code approved factors influencing gas pipe sizing:
- Maintaining sufficient gas pressure at the inlet of each appliance
- Keeping the minimum gas pressure at 5 in. w.c. for appliances which require the same, or almost the same, minimum pressure at the appliance inlet
- Satisfying the greatest inlet pressure requirement if the minimum pressure varies at each appliance or inlet
- Satisfying the minimum gas pressure requirement of the farthest appliance in the system
- Avoidance of exceeding the pressure rating of the appliance regulator in systems where pressure is greater at the point of supply. This does not apply to small systems with the source pressure of ½ psi

Underground gas piping installation

The following are the National Fuel Gas Code requirements underground gas piping installation:
- In order to avoid contact with underground structures, allow timely maintenance, and protect against damages from proximity of other structure, it is necessary to provide sufficient distance between gas piping and other underground structures
- Plastic gas piping should be insulated and protected from the heat
- Gas piping should be buried deep enough to prevent damage from pressure such as heavy traffic, unstable soil conditions, and settling of foundations

- Underground piping should have at least 12 inches of cover. In areas where damage is anticipated, the cover should be increased to 18 inches. If it is impossible to provide a 12-inch cover, piping should be installed in conduits or shielded
- Piping should be protected against corrosion in areas where it comes in contact with earth

Residential buildings

The following are the steps to install gas piping in residential buildings:
- Make sure that the gas service pipe and meters are installed by the gas utility company
- Use black pipe and galvanized fittings, or aluminum connections, if required extend a ¾-inch or 1-inch pipe to the gas range in the kitchen as directly as possible
- Place a ½-inch outlet for the gas water heater
- In large buildings, extend the pipe for the pilot light for oil burners in the boiler
- After installation, make sure the gas piping is free of dips or pockets where condensation may accumulate
- Place cleanouts where necessary
- Test the system for leaks using one of the suitable tests

Sizing considerations

To determine the gas pipe sizing, the following considerations are outlined by the National Fuel Gas Code:
- Allowable loss of pressure from point of delivery to appliance – minimum and maximum loss of pressure that may occur between the source of gas and the farthest appliance
- Maximum gas pressure requirement – appliance that demands the greatest pressure should be considered to determine the sizing
- Length of piping and amount of fittings – the total of all pipes and fittings
- Specific gravity of the gas served
- Diversity factor
- For special gas piping systems and appliances, as well as for conditions, such as longer runs, greater gas demands, greater pressure drops, or other circumstances not covered by the tables in the National Fuel Gas Code, pipe sizing should be determined on the basis of standard engineering practices approved by the proper authorities

Plumbing fixtures and piping holes

In most construction projects, the holes for plumbing pipes and other equipment are required to be made in advance. Therefore, if a plumbing specialist is working on such project, he/she should be able to locate these holes. The blueprint should be consulted first in order to determine the location of plumbing fixtures and piping.

Then, the plumbing specialist should mark on the blueprint where the holes must be made for each fixture and pipe. Holes 2 9/16 inches in diameter are marked in the walls for such fixtures as lavatories, kitchen sinks, bathtubs, etc. The same size holes should be marked for vents. For vents that run through the roof, 3-inch holes should be marked on the ceiling. Holes of 5 inches in diameter should be marked on the floor for floor drains.

Bushing

The main function of a bushing is to connect the male end of a pipe to a fitting of a larger size. A bushing can be a reducing or enlarging fitting. The bushing consists of a hollow plug with male and female threads to fit different diameters. Bushings used to reduce to one size should be made of malleable iron; bushings used to reduce to two or more sizes should be made of cast iron. A regular pattern bushing has a hexagon nut at the female end. A faced bushing has a faced end and is used with a long screw pipe and a faced lock nut to form a tight joint or to receive a male end fitting for close work.

Lead wall shields

Lead wall shields and a lag screw are used to fasten pipe hangers to concrete walls and floors. The lead shields are usually preassembled into a self-contained single unit. The shell-like unit has tapered internal threads for a portion of its length. The outside of the anchor has a series of circumferential ribs starting at the bottom and running for a major portion of its length. The back end of the anchor has two equally spaced ribs that extend beyond its diameter and run for a portion of its length. The short lead shields are used for anchoring in high grade concrete or where thickness of base material prohibits the use of a longer length shield. The long lead shields are used for lower grade base material or where extra anchoring strength is required.

Installing pipe through concrete

Usually, in new work in reinforced concrete buildings, the holes for piping should be sleeved in advance and the inserts should be placed for pipe hangers and supports. However, changes and additions require drilling of new holes. The following tools are used for drilling holes to install pipe through concrete walls, ceilings, or floors:
- A ⅜-inch hammer drill – used for drilling small diameter holes for pipe anchors; produces clean holes faster than a drill motor
- A depth guide rod – used on the side of the drill for drilling anchor holes; helps to drill to a required depth
- A 2-inch rotary hammer – used to drill holes up to 1⅛ inch in diameter with solid drill bits, and up to 2 inches in diameter with core type drill bits
- A wet core drilling outfit – used to drill holes in steel reinforced concrete walls

Pipe hanger installation

The following are the guidelines for pipe hanger installation through metal floor:
- If the pipe is installed through a metal deck with concrete fill, all inserts should be set before concrete is poured
- The inserts should be imbedded in the cured slab
- The hanger rods and clevis should be placed to make a complete installation
- Pipe clamps should be secured around the pipe with two bolts and rest on the floor slab
- Fire-rated sleeves may be used to prevent exposure to fire
- Watertight sleeves should be used on sections of piping that is passing through foundation walls and floor slabs on grade and is exposed to the outside elements

Installing pipe through wood floors

The plumbing specialist has to cut holes to install piping in the building's wood frame after the location of the piping and plumbing fixtures has been determined. The following are the tools used for cutting holes to install pipe through wood floors:
- A ½-inch right-angle drill – fits between a 16-ich stud or joist space; minimum size drill bits can be used to minimize the loss of structural strength
- An electric hacksaw – used for cutting holes for the pipe exceeding 2 inches in diameter and for cutting openings where the angles of pipe installation make it impossible to drill a hole
- A ¼-inch drill motor – used to drill ¼- to 1 ½-inch holes for water pipes passing through the wood structure of the building
- A keyhole saw – used for cutting openings in places where an electric saw cannot be used

Fire standpipe systems

The fire standpipe system is the most common fire protection system that consists of a water supply serving one or more hose outlets. A standpipe system is required for installation in buildings which are more than three stories high or whose area is more than 10,000 square feet. The locations of the standpipe systems are determined by the local codes. Standpipe systems can be of two types: wet and dry. In wet standpipe systems the piping is filled with water that is discharged under pressure through the opening in any hose valve. In dry standpipe systems, the piping is filled with air. Reduction of air pressure, which is regulated by a remote control system, activates the entrance of water into the system.

Safe pipe installation

The following are some of the general requirements for safe pipe installation:
- After installing a new system or remodeling the existing one, make sure to leave the building in a safe structural condition
- Meet the limitations for cutting, notching, and boring framing members to receive piping set by the local building and plumbing codes
- When installing pipes that penetrate fire-resistant floors, walls, or ceiling, make sure to protect the assemblies according to the local building code
- When trenching for piping that is required to run parallel to the footings of the building, do not extend the trench bottom below the 45° bearing plane of the footing or the wall

Sprinkler systems

Sprinkler systems consist of piping, valves, and sprinkler heads connected to one or more sources of water.

Five types of sprinkler systems
The following are the five types of sprinkler systems:
1. Wet pipe – automatic sprinklers are attached to water piping where water is under pressure at all times. The heat melts a fusible link and water discharges immediately onto the fire
2. Dry pipe – used in unheated areas. Automatic sprinklers are attached to piping carrying air under pressure. When a sprinkler head opens, the air is exhausted and water begins to discharge
3. Pre-action – prevents accidental water discharge from defective sprinkler heads. The piping is filled with air kept under low pressure. If there is a leak, an alarm is triggered, but the main valve does not open. The valve is activated by an automatic fire detection system
4. Deluge – sprinklers remain open; when the main valve is activated, water floods the affected area
5. Firecycle – heat detectors and electrical control panel regulate the flow of water. The system is used in locations where water damage must be minimized

Selection
Selection of the sprinkler systems depends on the type of the building and its hazard classification. Most buildings open for public are classified as light hazard; for example, hospitals, schools, nursing homes, offices, restaurants, churches, residences, etc. Most of the industrial buildings are classified as ordinary hazard; for example, automotive garages, warehouses, electric generating stations, cement plants, garment factories, etc. Facilities handling flammable materials are classified as extra hazard; for example, chemical plants, oil refineries, paint shops, etc.

Supporting horizontal piping

The following are the requirements for supporting horizontal piping of various types:
- Hub-type cast iron soil pipe – should be supported at 5-foot intervals if 10-foot pipe lengths are not used
- No-hub type cast iron soil pipe – should be supported at every other joint. If the developed length between hangers exceeds 4 feet, hangers should be provided at each joint
- Threaded iron pipe – should be supported at 12-foot intervals
- Copper tubing with the diameter of 1¼ inches or smaller – should be supported at 6-foot intervals
- Copper tubing with the diameter of 1½ inches or larger – should be supported at 10-foot intervals
- Plastic pipe conveying cold water wastes – should be supported at 12-foot intervals
- Plastic pipe conveying hot water wastes – should be supported for its entire length

Supporting vertical piping

The following are the requirements for supporting vertical piping of various types:
- Hub-type and no-hub type cast iron soil pipe – should be supported at the base and at each story of the building
- Rubber gasket joint soil pipe – should be supported at 5-foot intervals if 10-foot pipe lengths are not used
- Threaded iron pipe – should be supported at every other story of the building
- Copper tubing – should be supported at each story of the building
- Exposed and concealed plastic pipe with the diameter of 1½ inch or smaller – should be supported at 4-foot intervals
- Exposed and concealed plastic pipe with the diameter of 2 inches or larger – should be supported at each story of the building

Supporting horizontal and vertical glass piping

The following are the most commonly used methods of supporting pipe:
- Pipe strap – a metal strap attached to walls or ceilings with screws
- Perforated iron strap – used to support small water and heating pipes
- Clevis hanger – used for supporting large pipes; can be adjusted to line up the pipe
- F&M hanger – made of two malleable iron halves that are hinged at the bottom and bolted at the top
- Coil hanger – used to support ceiling coils or several parallel pipes

- 57 -

- Spring cushion – used for buildings with significant vibration, for example, in earthquake-prone areas
- Split ring hanger – used in commercial structures; can be disassembled while remaining attached to joists

<u>Requirements</u>
The following are the requirements for supporting horizontal and vertical glass piping:
- There should be only one anchor point in a line of glass pipe unless an expansion loop or swing joint is provided
- Anchors or hangers should be lined with sheet rubber or asbestos to prevent scratching of the pipe
- Hangers should be placed 12 inches from each end of a 10-foot length of pipe
- Hangers should permit the pipe to move 1 inch lengthwise or sidewise
- Horizontal branches should be supported every 8 feet
- Vertical branches should be supported at the base with a padded saddle support or beam clamps under a flange

Wood screws

Wood screws are used to fasten large hangers in place and to install wall carriers for various plumbing fixtures. Wood screws require a lot of effort to drive them in place manually. The screwdriver used to drive wood screws must be in good condition. It means that the blade should be square with the shank and as thick and wide as the slot head screw in which it is going to be used. Before driving a wood screw, use a drill bit in a drill motor to make a pilot hole a fraction of an inch smaller in diameter than the wood screw. This helps with driving the screw and prevents the base wood from cracking. A screw-driving bit can be obtained to go into a carpenter's brace. This allows leverage to be applied to the screw head on larger size screws.

<u>Pipe attached to wood</u>
The following are the most commonly used methods of attaching horizontal and vertical lines of pipe to wood surface:
- Reznor hook – a bent piece of wire with two pointed ends that are driven into a wooden joint. There are several varieties of Reznor hooks; for example, a U-hook, a J-hook, etc. The difference between different varieties is in the shape of the hook and in the method of attaching it to the wood surface
- Tube strap – can be made of copper, galvanized iron, or plastic; come in various sizes, from ½ and ¾ inch to 2 inches
- Perforated band iron – a thin band of iron perforated at equal intervals; can be cut into different lengths to fit the pipe

Types of insulation

The following are the most commonly used types of insulation: fiberglass, cellular glass, rock wool, polyurethane foam, closed cell polyethylene, flexible elastometric, rigid calcium silicate, phenolic foam, or rigid urethane.

Fiberglass insulation

Fiberglass insulation is the most preferred type of insulation. It comes with a vapor barrier jacket made of aluminum foil between layers of Kraft paper. Fiberglass insulation conductivity should not exceed 0.22 btu per square inch per hour at an average temperature of 75°F. For piping exposed to freezing conditions, thickness of the insulation for hot, cold, recirculation, and storm lines should be ½ and 1½ inch. Fittings and valves should be insulated with hydraulic setting cement of the same thickness as the insulation for piping.

Water line insulation

The following are the general recommendations for water line insulation:
- Before the cold season starts, drain water from swimming pool and water sprinkler supply lines according to the manufacturer's instructions. Do not put antifreeze in these lines unless directed
- Remove, drain and store hoses used outdoors. Close inside valves supplying outdoor hose bibs. Open the outside hose bibs to allow water to drain. Keep the outside valve open so that expanded water would not break the pipe
- Insulate all exposed hot and cold water pipes with such products as "pipe sleeve," "heat tape," "heat cable" or similar materials
- Carefully wrap the pipes, with ends butted tightly and joints wrapped with tape. Follow manufacturer's recommendations for using insulation products

Cleanouts locations

The cleanouts can be located in the following places:
- The main front cleanout can be found at the point of connection of the building sewer and drain at the outside wall of the building
- Stack base cleanouts can be found at the base of all vertical soil or waste stacks
- Bend cleanouts can be found at all turns where the change in direction is 90 degrees
- Other cleanouts can be found at the upper terminal of all horizontal branch drains
- On horizontal drainage piping which has a diameter of 3 inches or smaller cleanouts can be found at every 50 feet
- On horizontal drainage piping which has a diameter of 4 inches or larger cleanouts can be found at every 100 feet

Cleanout installation

The main front cleanout should always be installed at the outside wall of the building where the building sewer and the drain connect. This cleanout should be a full wye fitting placed in the direction of the flow. It can be installed inside or outside of the building. The main front cleanout should be positioned at least 2 inches above the floor or grade level to avoid removing of the cover and using the cleanout opening as a drain. However, in areas of traffic the cleanout should be placed level with the floor. Stack base should be installed at least 6 inches above the floor to prevent their use as drains. Installation of the cleanouts at the upper terminal of all horizontal branch drains might not be necessary if there is a plumbing fixture trap that can be removed and used as a cleanout. Cleanouts must always be accessible.

Piping insulation

Piping insulation should be in accordance with ASTM C547. Fittings, flanges, and valves (except valve stems, hand-wheels, and operators) should be insulated with pre-molded insulation of equivalent thickness and of same composition as insulation installed on adjacent piping.

Unicellular insulation should be used only on run-outs serving individual terminal units not exceeding 12', piping on packaged equipment, and piping exterior to a building. Vapor barrier jackets should be provided on pipe insulation, except on hot water lines or unicellular insulation.

Piping in trenches

The following are the steps for installing piping in trenches:
- Prepare the bottom of the trench by removing rocks, lumps, and other debris
- Carefully lower the pipe into the trench
- Dig holes for each coupling or bell to install the pipe
- Check the drainage pipe to make sure it has the proper grade. Raise or lower the pipe if necessary
- Make joints
- Test the pipe line and have it inspected by proper authorities
- Hand-backfill and tamp the dirt around the centerline of the pipe filling voids to prevent the pipe from holding the weight of the dirt
- Hand-backfill and tamp the dirt filling to at least 1 foot above the pipe
- Fill the trench with heavy material, such as lumps of dirt and rocks, using machinery

Footing drain installation

On construction sites where surface water accumulates during heavy rainstorms sub-soil drainage systems are required to protect the foundation wall and its footing supports from weakening and collapsing. The following are the requirements for footing drain installation:

- The drainage line should be installed slightly above the bottom of the footing
- The pipe should be laid on a bed of crushed rock that provides firm support. The pipe should also be covered with crushed rock for easy drainage
- Clay or concrete pipe, open-jointed, perforated, or horizontally split to facilitate seepage should be used
- Bituminous paper saturated at the openings to prevent sand from entering the drain should be used
- The minimum required slope for footing drain is ½ inch per foot

Rocky trenches

The following code requirements should be observed when pipe is installed in rocky trenches:

- Remove the rock to the depth of 3 inches below the installation level of the bottom of the pipe
- Provide uniform load-bearing support for the barrel of the pipe between joints using the fine backfill materials
- Carefully tamp the materials into place up to the installation level of the pipe
- Make sure to select backfill material that is free of rock, broken cement, frozen lumps, and other debris
- Bring the backfill up evenly on each side of the pipe to keep it aligned
- Compact the backfill under the pipe and at the sides

Calculating drainpipe grade

The grade should be such that the water running through the pipe should have a velocity of 260 feet per minute. The following is the formula for calculating the grade:

$$F = \frac{\frac{L}{10d}}{}$$

F=total fall or grade in feet
L=length in feet
D=diameter of the pipe

The following example illustrates the calculation of the slope with this formula:
- The diameter of the 8-foot sink waste is 2 inches
- Find the slope to give the pipe a velocity of 260 feet per minute
- 2" x 10 = 20
- 8' ÷ 20 = 0.4'
- 0.4' x 12" = 4.8" or 4 3/4"
- 4 3/4" ÷ 8' = 5/8"
- The slope should be 5/8 inch per foot.

Grading

Grade is an amount of slope or fall of a pipe in relation to a horizontal plane. Grading ensures that the pipe is self-scouring and the sewage is disposed of properly. The grade of drainage pipes should be between ¼ and ½ inch per foot. Below are the steps for grading:
- Determine the grade using the size of the pipe and the length of the drain to reach the flow rate of 260 feet per minute
- Use a level and a block of wood representing the grade. If the length of the level is 2 feet and the grade is ¼ per foot, the thickness of the block should be ½ inch
- Place the level on the pipe and a block of wood under the low end of the level
- When the air bubble is in the center of the glass chamber, the grade is proper

Roof vent terminal installation

Vent terminals are extended through the roof to vent the sewer in the atmosphere around the building. The terminals should be extended at least 1 foot to prevent rainwater and foreign materials, such as leaves and insects, from falling into the vent and blocking it. Vent terminals should be installed at least 10 feet away from doors, windows, or other ventilating openings to avoid foul-smelling odors of the sewer.

If it is impossible to keep such distance, vent terminals should be installed at least 2 feet above ventilating openings. In cases where roofs are used as sundecks, vent terminals should be extended at least 7 feet above the roof. It is necessary to seal vent terminals to the roof to prevent leakage. For vent terminals installed in areas with cold climates 2-inch pipes should be used.

Roof flashings

Roof flashings serve as watertight protection of vent pipes. Roof flashings can be adjusted to a variety of angles to fit any roof, flat or pitched. To adjust the angle of the roof flashing, it is necessary to rotate the body of the flashing and tap the neoprene or lead ring tightly against the pipe.

The most commonly used materials for roof flashings are sheet steel and copper. If the flashing is installed on a pitched roof, its upper part should be slipped under the roofing material while its lower part is placed over the roofing material. Flexible roofing cement is applied to all the components of the installation: the seal, the pipe, the nails, and around the edges of the flashing.

Backing boards

Backing boards, or supports and hangers, are devices used for supporting and securing pipe, fixtures, and equipment to walls, ceiling, floors, and other structural members. For example, backing boards are installed to support lavatories, bathtubs, water closets, etc. The best material for fabricating a backing board is wood, such as pine shelving material. Hard wood or wood filled with knots should be avoided because it tends to split when nails or screws are driven into it.

To install a backing board, follow these steps:
- Determine the height of the board and cut the appropriate length
- Cut the cleats and align them on the studs
- Nail the cleats in place
- Nail the board to the studs

Bathtubs
A backing board should be installed for recessed bathtubs to secure the rear lip of the tub. The board can be horizontal, 2 by 3 inches, and run the entire length of the tub. Legs are required for the horizontal wooden support to avoid pressing the weight of the tub on the nails. Vertical support may also be provided. Two wooden legs, measured 2 by 3 inches or 3 by 4 inches can be placed on the floor and nailed to the studs.

Water closets
If a water closet is intended for installation in a room not originally designed for it, a joint may interfere with the outlet to the fixture. The joint should be boxed in to protect the flooring from weakening.

Lavatories
All wall-hung lavatories should be supported with a backing board. The following are the steps for installing a lavatory backing board:
- A backing board should be provided behind the wall before the wall covering is installed
- The mounting bracket of the lavatory should be attached to the backing board through the wall covering
- The lavatory should be hung upon the bracket

If a lavatory is of pedestal type, the backing board is not required.

<u>Shower heads</u>
The backing board for the shower head should be installed no less than 6 feet 6 inches above the floor. The following are the steps for installation of backing boards for shower heads:
- The showerhead arm is threaded into a drop-eared elbow behind the wall
- The elbow is attached to the board with two screw hole lugs

Loop vent installation

Loop and circuit vents are used on a line of fixtures. Loop vents are installed on single-story buildings or on the top floors of the multi-story buildings. In loop vents, the vent branch goes back into the stack vent at a point located above the drain inlets. The horizontal branch vent pipe should pitch back to the traps. The branch drain should continue full size to the last fixture connection.

Circuit vent installation

The circuit vent is limited in size; the total pitch of the drain should not exceed 18 inches. The size limitation of the pipe determines the number of fixtures that can be connected. A fixture should be connected at the end of the vertical vent of the drain in order to wash it out. A cleanout plug should also be placed at the end of the drain.

Adjustable horizontal carrier system

An adjustable horizontal carrier system is most suitable for institutional or commercial applications as it allows the installations of multiple water closets in batteries. The system should be installed in accordance with ANSI Standard A.112.19.2, Vitreous China Plumbing Fixtures, Paragraph 5.1.3.3. This standard accommodates the installation of a "Physically Handicapped" type siphon jet water closet that requires a dimension of 18" from the floor to the rim of the bowl.

Standard water closets should be installed with a 15" floor to rim installation with a 5½-inch centerline of closet outlet to finished floor dimension with a ⅛ inch pitch to the run. The wheelchair closet should have an 8½-inch finished floor to centerline dimension.

Follow these steps to install the carrier:
- Install carrier fitting in waste line rough-in. Support at the correct pitch and height
- Install faceplate and foot assembly using the faceplate gasket to seal between the faceplate and the fitting
- Secure body and faceplate together with machine screws and washers

Concealed arm carriers

Follow these steps to install a concealed arm lavatory carrier:
- Locate position of the carrier to plan, using lag bolts
- Secure fixture support uprights to the unfinished floor slab with ½-inch lag bolts
- Make sure the center of the concealed arm nipple is 2¼ inches offset from the centerline of the upright member
- Set brackets at proper height, install nipple in brackets and threaded crossbar, and install arms in nipples
- Take the lavatory dimensions and measure the thickness of the wall including finish. Make adjustments as necessary to ensure that the lavatory fits over arms and that the leveling and locking devices are accessible through the fixture openings

Bathroom fixtures

The following are the general guidelines for installing bathroom fixtures:
- Uncrate and inspect the fixture and its trim for defects or damage
- Check the measurements for the waste and water supply piping with the rough-in drawing provided with the fixture
- Attach the wall or floor supports for the fixture
- If necessary, attach the fixture trim
- Hang or set the fixture
- Align, level, and plumb the fixture if necessary
- Secure the fixture fasteners
- Connect the fixture to the water supply and waste piping
- Purge the water supply piping to release air and remove dirt from the piping
- Test the water supply and waste connections
- Join the fixture to the wall with caulk or grout
- Clean and inspect the fixture

Plumbing fixtures installation phases

The plumbing fixtures installation takes place in two phases: rough-in and finishing.

At the rough-in phase all parts of the plumbing system that can be completed before the installation of fixtures are installed. The rough-in installation includes installation of drainage piping, vent piping, water supply piping, fixture supports. Plumbing fixtures come with rough-in drawings provided by manufacturers. The drawings assist the plumbing specialists in proper and safe installation of the piping to fit fixtures and appliances.

Finishing is the actual installation of fixtures and appliances. It happens after the rooms are completed and are ready to be occupied. Built-in fixtures, such as a bathtub, shower base, etc., are exceptions to this rule.

Shower installation

The following are the International and State Plumbing Codes requirements for shower installation:
- Showers should be equipped with a shower base of a minimum area of 900 square inches that should extend no less than 30 inches in any direction measured from the inside
- If the minimum area of a shower base required is 1,024, the minimum size requirements should be maintained for a vertical height equal to 70 inches above the drain. Only grab bars, faucets, and showerheads may be protruded into this area
- The wall enclosure must be waterproof and extend from the finished floor to at least 6 feet
- The wall enclosure must extend at least 70 inches above the drain

Bathtub and shower bases

The bathtub and shower bases are built-in fixtures, so it is required that they should be installed before the plasterboard and ceramic tiles are placed. The installation of the sanitary drainage and vent piping must be complete before installation of the bathtub and shower bases.

To install the shower base, it is necessary to clean the floor for even rest; set the base precisely into its location; and seal the connection to the 2-inch pipe inlet of the P-trap with a caulked lead and oakum joint.

To install the bathtub base, it is necessary to attach a piece of board along the back wall of the tub space to support the back edge of the bathtub; set the base precisely into its location; attach the waste and overflow fitting; and connect the P-trap to the bathtub drain.

Trip lever mechanism

The following are the steps for adjusting the trip lever mechanism:
- Remove the screws holding the overflow plate
- Pull the entire linkage forward through the hole in the tub
- Loosen the locknut at the base of the yoke
- Turn the threaded rod clockwise to raise the plunger and counterclockwise to lower the plunger
- After making the adjustment, tighten the locknut

It might take several adjustments to get the correct linkage length to make sure that the plunger will seal and open correctly. If the linkage cannot be adjusted to stop the flow of water due to wear or damage, replace the trip lever mechanism.

Lavatory faucet

Installation of a lavatory facet gives the fixture a finished appearance. To install a lavatory facet, follow these steps:
- Connect the faucet to the water supply line by joining the threaded portion of the shank, a coupling nut, and the short supply pipe
- Make sure the supply pipe uses a flexible composition or soft-metal packing and a friction ring to provide a leak-proof connection with the faucet
- Make a waste connection to the lavatory bowl with a plug top fitting which consists of a short length of pipe with a flange at one end and threads at the other end
- Use the plain plug top for lavatories with a separate overflow outlet; use the ported plug top for lavatories with a combined overflow and waste outlet

Sump tanks

The drainage must flow into the sump by gravity. Locate the sump tanks for future inspection, maintenance, and repair. The following are the requirements for sump tanks:
- The sump tank should have a minimum diameter of 18 inches, and a minimum depth of 24 inches
- The sump tank should be made of approved hermetic material, such as tile, steel, concrete, or plastic.
- The bottom of the tank should be solid and provide enough support for the pump
- The tank should be properly ventilated and equipped with a removable gas-tight cover
- Metal tanks should be protected from corrosion
- In drainage systems intended for public use, duplex pumping equipment is required

Clothes dryer vents

A clothes dryer vent exhaust must always lead outside. To install a dryer exhaust from a basement, two 90° bends and 10 to 15 feet of flexible plastic pipe are necessary. Vent caps should have a 4-inch opening or the mini-louver doors. These types offer the least airflow resistance. The wall dryer vents should not be used on a roof. A roof vent cap designed to shed rainwater should be used. Roof clothes dryer vent installation is similar to basic roof vent installation. The vent should not be visible from the ground. Special large clamps should be used to secure the pipe to the fittings to avoid lint accumulation. The seam between metal pipe and fittings should be taped with the foil faced duct tape. If the dryer is vented through the roof, hidden sections of metal pipe must be insulated with at least three inches of fiberglass.

Sink installation

The following are the International and State Plumbing Codes requirements for sink installation:
- Sinks should be equipped with drains of a minimum diameter of 1½ inch
- Strainers or crossbars, which prevent foreign objects from falling into the drain, should be installed
- If a sink has a garbage disposer, the sink drain opening should a minimum diameter of 3½ inches
- Garbage disposers should be equipped with a drain of at least 1½ inch and a trap
- A drainage connection made with removable slip-nuts and washers should be accessible

Sump tanks pipe

To install piping for sump tanks, follow these steps:
- Install a swing check valve and a gate valve in each of the twin discharge lines between the pump and the building gravity system. The check valve is required only for one- or two-family house systems
- Connect the sump discharge pipe into the top of the building gravity drain with a wye fitting
- Make this connection at least 10 feet from the base of any soil or waste stack
- Make sure that the minimum capacity of a sewage ejector depends on the required diameter of the discharge pipe with the exception of grinder ejectors that receive water closet discharge

Air gaps in dishwasher installation

Dishwashers must be equipped with either a gap flow protector or an air gap installed on the water-supply pipe to prevent back-siphonage. The air gap should be mounted on the countertop or in the rim of the kitchen sink. The air gap works in the following way: it forces the waste discharge of the dishwasher through the atmosphere and down through a separate discharge hose. This eliminates the possibility of back-siphonage or a backup from the drainage system into the dishwasher.

Different codes have different requirements for dishwasher drainage. Some codes require that the dishwasher drainage should be trapped separately and vented, or discharged indirectly into a properly trapped and vented fixture.

Other codes require that the discharge hose should enter the drainage system through an individual trap, a trapped fixture, or through a wye connection in the kitchen sink drainage.

<u>Domestic dishwasher</u>
A domestic dishwasher is rated at 2 dfu of waste discharge and requires a 1½-inch waste pipe and a 1¼-inch vent pipe if it drains by gravity into the drainage system. However, in most of domestic dishwashers the waste is pumped out.

The pump-out waste should be connected with a rubber hose or copper tubing to a dishwasher tailpiece beneath the kitchen sink basket strainer or into the dishwasher drain connection on the garbage disposal. In both types of installation, it is necessary to make a high loop in the pump discharge piping as it connects the dishwasher and the drain. The loop allows the dishwasher to drain properly and prevents backflow from the kitchen sink and/or garbage disposal into the dishwasher.

Garbage disposal

A garbage disposal, or a food waste disposer, is an electric device that grinds food wastes into a pulp and discharges this waste into the drainage system. Water is necessary for a garbage disposal to operate properly. A typical domestic garbage disposal is mounted beneath one of the compartments of a kitchen sink in place of a basket strainer assembly.

A garbage disposal may discharge its waste into a P-trap or into a continuous waste fitting. If the disposal discharges into a continuous waste fitting, the disposal waste might backflow into the other sink compartment. To prevent this, the tee connection joining the disposal waste and the waste for the other sink compartment must contain an internal baffle.

<u>Installation</u>
The following are the steps to install a garbage disposal:
- Remove the waste pipes from the sink strainer to the threaded fitting at the wall or floor
- Remove the sink strainer. Install the flange that comes with the disposer applying plumber's putty under its lip to form a seal with the sink bowl
- Slip the mounting assembly gasket and mounting and retaining rings over the neck of the sink flange. Tighten the screws. Remove excess putty
- Mount the disposer. Attach the drain elbow. Slip the disposer's slotted flange over the mounting bolts

- Connect a two-piece tubular P-trap to the drain elbow and the drain fitting at the wall. Cut the P-trap and rotate the trap section of the-P trap and the disposer. Tighten the waste pipe fittings and the disposer
- Follow the manufacturer's instructions to make the power connection

Automatic gas water heater

An automatic gas water heater consists of a vertical storage tank encased in sheet metal insulated to reduce heat loss. A pre-heater is screwed into the bottom of the tank. The flame is directed against the pre-heater. The bottom of the tank as well as the flue absorbs the heat.

The gas is controlled by an automatic thermostat that is installed in the side of the tank to allow the incoming cold water to turn on the gas. A draw-off cock is placed at the bottom and a draft hood is installed on the flue pipe to prevent down draft. The flue is connected to the chimney and allows the gas fumes to escape. Some automatic water heaters do not have pre-heaters; rather, the whole tank acts as a heating surface.

Thermostat
A thermostat is installed inside the water heater to control the gas valve. A thermostat consists of a carbon rod encapsulated in a copper tube closed at the end. While the water in the tank is hot, the copper tube expands causing the spring to close the gas valve. When the hot water is drawn and the cold water flows around the copper tubing, the metal contracts causing the carbon rod to push against the valve stem and lift the valve from its seat. The valve opens allowing the gas to flow to the burner.

Thermostats are usually equipped with a temperature indicator so that the water temperature can be easily monitored.

"Instant Hot" water heaters

Advantages of "Instant Hot" water heaters:
- Electric units can be installed at the point of use
- Energy costs are low
- Heaters are easy to install
- Electric units do not require a temperature and pressure valve or a tank.
- Heaters are environmentally friendly

Disadvantages of "Instant Hot" water heaters:
- They need a minimum flow rate and pressure to turn on. The faster water flows through them, the lower the temperature rise
- Electric units need heavy gauge wire. For example, the 9.5 kw must have 8 ga wire and a 50 amp breaker
- Gas units need a larger flue pipe and larger gas supply than a conventional water heater

Electric water heater

An electric water heater is equipped with a storage tank large enough to hold 24-hour water supply. On top of the tank, there is a coldwater inlet and a hot water outlet. From the coldwater inlet, the water is delivered to the bottom of the tank where it is heated. The warm water rises to the top of the tank, escapes through the hot water outlet, and is distributed throughout the building.

The relief valve is installed either on the side of the tank or in a tee close to the hot water outlet. There are two heating elements, lower and upper. The lower heating element maintains the standby temperature of the tank, while the upper heating element heats water at the top for immediate use. The tank is insulated to prevent heat loss.

Water heater installation

The following are the International and Plumbing Codes requirements for water heater installation:
- When a water heater with an ignition source is installed in a garage, an elevated base must be provided. The base should be at least 18 inches above the garage floor
- Solid-, liquid-, or gas fuel-fired water heaters should not be installed in rooms that contain air at a pressure greater than that of the outside atmosphere and contain air handling machinery simultaneously
- Fuel-fired water heaters must not be installed in bedrooms, bathrooms, or closets, unless equipped with a direct-vent system
- In areas prone to earthquakes, water heaters must be equipped with supports whose design allows them to withstand the seismic forces indicated in the International Building Code
- The minimum measurements of the exit route for an attic water heater should be 30 inches high, 22 inches wide, and 20 feet long

Indirect water heater

An indirect water heater consists of a copper coil with a cast-iron jacket. It is connected to a steam boiler and is placed below the waterline. The water from the steam boiler circulates through the indirect heater and around the copper coil. A storage tank is connected to the steam boiler; the water runs through the heater and into the tank. The inlet is run to the boiler outlets 2 inches below the waterline.

These are the requirements for installation of indirect heaters:
- Gate valves should be installed between the heater and the boiler
- The bottom of the tank should be placed at least 12 inches above the heater coil for gravity circulation
- The upper connection of the storage tank should be hooked to the upper connection of the coil
- The lower outlet of the tank should be connected to the lower connection of the coil

Combustion air openings size

The following are the International and Plumbing Codes requirements for size of combustion air openings for gas- or liquid-fired water heaters in buildings of ordinary tightness:
- Water heaters located in confined space using air from inside building should have two openings into enclosure each having 1 square inch per 1,000Btu/hr input freely communicating with other unconfined interior spaces
- Water heaters located in confined space using part of air from inside building should have two openings into enclosure from other freely communicating unconfined interior spaces each having an area of 100 square inches plus one duct opening to outdoors with an area of 1 square inch per 5,000Btu/hr input
- Water heaters located in confined space using all air from outdoors should have two openings that conform to the requirements for appliances located in buildings with unusually tight construction

The International and Plumbing Codes requires these sizes of combustion air openings for gas- or liquid-fired water heaters in buildings of unusually tight construction:
- Water heaters in unconfined space should have two openings into enclosure, each having 1 square inch per 5,000Btu/hr input

Water heaters in confined space using part of air from outdoors or from space freely communicating outdoors should have the following:
- Two vertical ducts or plenums each having 1 square inch per 4,000Btu/hr input
- Two horizontal ducts or plenums each having 1 square inch per 2,000Btu/hr input
- Two openings in an exterior wall of the enclosure each having 1 square inch per 4,000Btu/hr input

- One ceiling opening and one vertical duct to ventilated attic, each having 1 square inch per 4,000Btu/hr input
- One opening in enclosure ceiling to ventilated attic and one in enclosure floor to ventilated crawl space, each having 1 square inch per 4,000Btu/hr input

Water heater connections

The following are the International and Plumbing Codes requirements for water heater connections:
- Cutoff valves should be installed on a cold-water branch line from a main water supply to a hot-water storage tank or water heater
- The cutoff valve should be accessible on the same floor and located near the equipment
- The cutoff valve must serve only the hot-water storage tank or water heater
- The cutoff valve must not interfere with or cause disruption of the cold-water supply to the remainder of the cold-water system
- Connections between a circulating water heater and a tank should facilitate the proper circulation of water through the water heater

Water heater installation in attics

The following are the International and Plumbing Codes requirements for installation of water heaters in attics:
- An opening and unobstructed passageway should be provided to allow the removal of the water heater. The opening should be large enough to fit the water heater, that is, at least 30 inches high, 22 inches wide, and no more than 20 feet long
- The exit area should be covered with solid continuous flooring at least 24 inches wide
- A level service area should be provided in front of the water heater. The service area should be at least 30 inches deep and 30 inches wide
- The service area should also be provided with a clear access opening. The minimum dimension of the access opening should be 20 by 30 inches

Boiler

The following are the fittings and devices necessary for proper operation of a boiler:
- A safety valve – should be installed to protect the boiler and its users in case of malfunction. The valve can be adjusted to pen and relieve internal pressure if it rises above the safe level. There is more than one safety valve installed on a typical boiler
- Water gauges – installed to provide a means of visually checking the level of water. Water gauges may be equipped with floats that activate an alarm in case water drops below the safe level

- Fusible plugs – installed to relieve pressure in case of malfunction
- Injectors – automatic devices installed to supply water to a boiler against the high interior pressure

Employee use ratios

According to the Uniform Plumbing Code, the following is the ratio for the employee use male and female water closets in penal institutions:
- 1-15 persons – 1 water closet
- 16-35 persons – 2 water closets
- 36-55 persons – 3 water closets
- Over 55 – add one more fixture for additional 40 persons

The following is the ratio for the employee use male urinals:
- 1-9 persons – 0 urinals
- 10-50 persons – 1 urinal
- Over 50 – add one more fixture for additional 50 males

The following is the ratio for the employee use lavatories:
- 1-40 males – 1 lavatory
- 1-40 females – 1 lavatory

The following is the ratio for the employee use drinking fountains:
- 150 persons – 1 drinking fountain

Restaurant, pub, and lounge ratios

According to the Uniform Plumbing Code, the following is the ratio for the restaurant, pub, and lounge male and female water closets:
- 1-50 persons – 1 water closet
- 51-150 persons – 2 water closets
- 151-300 persons – 3 water closets
- Over 300 – add one more fixture for additional 200 persons

The following is the ratio for male urinals:
- 1-150 persons – 1 urinal
- Over 150 – add one more fixture for additional 150 males

The following is the ratio for male and female lavatories:
- 1-150 persons – 1 lavatory
- 151-200 persons – 2 lavatories
- 201-400 persons – 3 lavatories
- Over 400 – add one more fixture for additional 400 persons

Cell and exercise room

According to the Uniform Plumbing Code, the following is the ratio for the cell and exercise room water closets in penal institutions:
- 1 water closet per cell
- 1 water closet per exercise room

The following is the ratio for the cell and exercise room male urinals:
- 1 male urinal per cell
- 1 male urinal per exercise room

The following is the ratio for the cell and exercise room lavatories:
- 1 lavatory per cell
- 1 lavatory per exercise room

The following is the ratio for the cell and exercise room drinking fountains:
- 1 drinking fountain per cell
- 1 drinking fountain per exercise room

Hospital room and ward ratios

According to the Uniform Plumbing Code, the following is the ratio for the hospital waiting room, individual room, and ward room water closets:
- 1 water closet per waiting room
- 1 water closet per individual room
- 1 water closet per eight patients in a ward room

The following is the ratio for the hospital waiting room, individual room, and ward room lavatories:
- 1 lavatory per waiting room
- 1 lavatory per individual room
- 1 lavatory per ten patients in a ward room

The following is the ratio for the hospital individual room and ward room bathtubs or showers:
- 1 bathtub or shower per individual room
- 1 bathtub or shower per twenty patients in a ward room

The following is the ratio for the hospital waiting room, individual room, and ward room drinking fountains:
- 1 drinking fountain per 150 persons

Hospital plumbing fixtures

According to the Uniform Plumbing Code, the following is the ratio for the employee use male and female water closets in hospitals:
- 1-15 persons – 1 water closet
- 16-35 persons – 2 water closets
- 36-55 persons – 3 water closets
- Over 55 – add one more fixture for additional 40 persons

The following is the ratio for the hospital employee use male urinals:
- 1-9 persons – 0 urinals
- 10-50 persons – 1 urinal
- Over 50 – add one more fixture for additional 50 males

The following is the ratio for the hospital employee use lavatories:
- 1 lavatory for additional 40 males
- 1 lavatory for additional 40 females

Elementary school student

According to the Uniform Plumbing Code, the following is the ratio for the student use male and female water closets in elementary schools:
- 1 water closet for each 30 male students
- 1 water closet for each 25 female students

The following is the ratio for the student use male urinals:
- 1 urinal for each 75 male students

The following is the ratio for the student use male and female lavatories:
- 1 lavatory for each 35 male students
- 1 lavatory for each 35 female students

The following is the ratio for the student use drinking fountains:
- 1 drinking fountain per 150 students;
- Over 150 students – add one more fixture for additional 150 students.

Nursery school student use

According to the Uniform Plumbing Code, the following is the ratio for the student use male and female water closets in nursery schools:
- 1-20 students – 1 water closet
- 21-50 students – 2 water closets
- Over 50 – add one more water closet for additional 50 students

The following is the ratio for the student use lavatories in nursery schools:
- 1-25 students – 1 lavatory
- 21-50 students – 2 lavatories
- Over 50 – add one more fixture for additional 50 students

The following is the ratio for the student use drinking fountains in nursery schools:
- 1 drinking fountain per 150 students
- Over 150 students – add one more fixture for additional 150 students

School employee use

According to the Uniform Plumbing Code, the following is the ratio for the employee use male and female water closets in schools:
- 1-15 persons – 1 water closet
- 16-35 persons – 2 water closets
- 36-55 persons – 3 water closets
- Over 55 – add one more fixture for additional 40 persons

The following is the ratio for the employee use male urinals:
- 1-50 persons – 1 urinal
- Over 50 – add one more fixture for additional 50 males

The following is the ratio for the employee use lavatories:
- 1-40 males – 1 lavatory
- 1-40 females – 1 lavatory
- Over 40 – add one more fixture for additional 40 persons

Secondary school student use

According to the Uniform Plumbing Code, the following is the ratio for the student use male and female water closets in secondary and other schools:
- 1 water closet for each 40 male students
- 1 water closet for each 30 female students

The following is the ratio for the student use male urinals:
- 1 urinal for each 35 male students

The following is the ratio for the student use male and female lavatories:
- 1 lavatory for each 40 male students
- 1 lavatory for each 40 female students

The following is the ratio for the student use drinking fountains:
- 1 drinking fountain per 150 students
- Over 150 students – add one more fixture for additional 150 students

School dormitory student use

According to the Uniform Plumbing Code, the following is the ratio for the school dormitory water closets:
- 1 water closet per ten males
- 1 water closet per eight females
- Add one fixture for additional 25 males and one for 20 females

The following is the ratio for male urinals:
- 1-25 males – 1 urinal
- Over 150 – add one fixture for additional 50 males

The following is the ratio for lavatories:
- 1 lavatory per twelve males
- 1 lavatory per twelve females
- Add one lavatory for additional 20 males and one for 15 females

The following is the ratio for bathtubs or showers:
- 1 fixture per eight males or females
- Add one fixture for additional 30 females
- Over 150 persons – add one fixture for additional 20 persons

The following is the ratio for drinking fountains:
- 1 fixture per 150 persons.

Clinical sink installation requirements

The following are the International and State Plumbing Codes requirements for clinical sink installation:
- Clinical sinks, which are also called bedpan washers, should be equipped with an integral trap which should be visible
- The contents of the sink should be removed by siphonic or blowout action
- The trap seal should be replenished automatically and every flushing of the sink should clean the sides of the fixture by a flush rim
- The connection of clinical sinks to the DWV system should be similar to water closet connection
- Clinical sinks in utility rooms may substitute service sinks
- Service sinks may not be replaced with clinical sinks

Sewage treatment installation

The following is the procedure for obtaining an approval for installation of private domestic sewage treatment and disposal systems in industrial buildings and facilities:

- Submit the complete plans and specifications of the industrial facilities, such as trailer parks, schools, hotels, apartment buildings, theaters, etc., to the local plumbing code administrative authority for approval
- Provide plans of the septic tank and sewage disposal system
- Identify the location of the water well and distances to lakes and streams, and to the property lot lines
- Measure lot size including grade or slope
- Perform soil boring and percolation tests
- Identify proposed use and occupancy of the buildings and facilities

Sterilizer installation requirements

The following are the International and State Plumbing Codes requirements for sterilizer installation:

- Both concealed and exposed piping for sterilizers should be accessible
- A gravity system should be provided for steam piping to a sterilizer to control condensation and prevent moisture from entering the fixture
- Sterilizers should be equipped with devices that control the steam vapors
- Sterilizer drains should be piped as indirect wastes
- Sterilizers must be equipped with leak detectors that should expose leaks and carry unsterile water away from the sterilizer
- Sterilizers should never be cleaned with acids or other chemical solutions while they are still connected to the plumbing system

Natural gas system installation

Natural gas systems serve water heaters, boilers, burners and other equipment. Gas service companies are responsible for installation of the gas service pipe and the gas meter.

Plumbing specialists may be required to install interior gas piping, which starts at the meter. For residential buildings, gas regulators are required for installation if the pressure exceeds ½ psi. For commercial and industrial buildings, pressures up to 3 psi are allowed. Schedule 40 black steel pipe with 150 lb. black malleable iron fittings is required for natural gas systems.

All components of the gas system are regulated by valves. Brass or iron-body valves are required for ½ - 2-inch pipes; lubricated plug valves are required for pipe sizes 2½" and more.

Public use requirements

The following are the requirements for public use plumbing fixtures:
- Plumbing fixtures should have smooth, nonabsorbent surface free from defects and damage
- Special fixtures may be made of soapstone, stoneware, nickel copper alloy, lead, and other materials
- Water closet bowls and traps should be made in one piece and in a form designed to hold enough water to the trap overflow to prevent contamination of surfaces when flushed
- Water closets should be of elongate shape with the open front seats made of smooth nonabsorbent materials
- Water closet tanks should have sufficient flushing capacity to facilitate proper flush connected water-closet bowls
- Floors should be made of nonabsorbent materials and provide at least 18 inches of space from front and on both sides of the closet, and extend at least 24 inches up the wall in back of the public water closet

Acid waste systems installation

Acid waste systems are installed in laboratory facilities at hospitals, schools, research institutions, etc. Acid waste systems are designed to convey waste containing harmful acids from laboratory sinks and other receptacles to the building sanitary drain. Neutralizing sump tanks filled with limestone are installed adjacent to each acid sink or fixture to dilute harmful wastes. In cases when one neutralizing sump tank is used for several fixtures, it should be located at the lowest story above the sanitary building drain. Deep seal traps should be provided for each acid fixture. All acid waste installations should be made in acid resistant piping, except for the outlet side of the neutralizing sump tank. The following materials are used for acid waste systems:
- Glass
- Polypropylene
- Cast iron high-silicone

Medical gas systems installation

The medical gas system in a hospital consists of the following subsystems:
- Oxygen – used for respiratory therapy; supplied through a storage tank or a gas manifold
- Nitrogen – used to drive surgical tools; supplied through a gas manifold
- Compressed air – used in intensive care units, emergency rooms, etc.; supplied by reciprocating air-cooled or oil-free air compressors
- Nitrous oxide – used for anesthesia; supplied through a gas manifold
- Vacuum – used in laboratories and surgical rooms; supplied through vacuum pumps

Medical gas systems are regulated by zone control valves that can shut off the supply of gas in each particular location. Specific outlets are required for different facilities. For example, in patient rooms, individual outlets are provided; in surgical suites, surgical ceiling columns are installed.

Backwater valve

Backwater valves are check valves installed in drainpipes to prevent the sewage from flowing back into the building. Backwater valves should be installed in the following cases:
- when fixtures are installed in basements
- when sewage is pumped up to a house drain
- when buildings are located near rivers

There are two types of backwater valves, the swing type and the balance type. The swing type backwater valve has a brass disk hinged at the top. The disk closes against the brass seat. The swing type backwater valve opens only when water flows in the right direction; it is closed for most of the time not allowing the proper ventilation of the drainage system. The balance type backwater valve has a disc on one end of an arm balanced in the center. An adjustable weight keeps the valve open under normal use.

All types of valves like any other plumbing equipment, fixture, and device come with the manufacturer's specifications. The specifications contain an isometric drawing of a valve identifying measurements, a list of required parts, nameplate data, and recommendations for installation. For example, the following are the manufacturer's recommendations for the installation of a pressure balancing valve:
- Product specifications

Rough-in valve includes the following:
- Integral isolation sops
- Diaphragm type pressure balancing cartridge
- Integral venture-twin el not necessary for installation with diverter tub spout
- Mechanical handle stop
- Comfort control/high temperature limit stop
- Minimum operating pressure – 20 psi
- Minimum operating pressure – 145 psi
- Flow rate at 45 psi – 4.5 gpm
- Recommended hot water temperature – 120-140°F

Installation requirements

The following are the requirements for backwater valve installation:

- Backwater valves should be installed in every location in the drainage lines where fixtures are subject to back pressure or backflow
- Bearing parts of backflow valves should be made of corrosion-resistant material and constructed to ensure a positive mechanical seal against backflow
- The diameter capacity of a backflow valve should be at least equal, but no less than the diameter of the pipe it is installed on
- Backflow valves can be installed in a floor drain, in soil waste line with extension cover, with offset for use in new construction, in existing drainage line, in a sewer terminal, or terminal end of a drainage line

Swimming pool and filtration installation

In projects requiring the installation of a swimming pool plumbing specialists are responsible for installation of pool filtration piping and equipment. The filtration system consists of the following components:

- Filter
- Circulating pump
- Pool heater (where required)
- Soda ash and hypochlorinator
- Alum feed
- Filtration pipe, fittings, and valves
- Gutter drain piping and fittings
- Main drain piping and fittings
- Pool supply piping and fittings

The most commonly used filters are sand filters. The pool pump ensures that the water from the swimming pool moves through the filter every day, removing pollutants and disinfected organic materials as quickly as possible. Diatomaceous earth, cartridge, and zeolite filters are also widely used.

Globe valve

A globe valve is a control valve installed on pipes carrying water, air, gas, oil, or steam. Globe valves can be fully opened or partially closed to regulate the flow. The most commonly used materials for manufacturing globe valves are brass, bronze, steel, and cast iron.

There are four types of discs that can be used to stop the flow in a global valve:
1. The conventional disc – closes against a beveled seat
2. The composition disc – can be used in pipes carrying hot or cold water, air, oil, and steam
3. The needle valve type – is used in pipes carrying gasoline or oil for fine throttling control
4. The plug-type valve – is used in pipes carrying steam; has a broad contact between the disc and the seat; made of special alloy steel

<u>Installation</u>
The following are the principles of globe valve installation:
- Screwed, welded, sweated, or flanged joints should be used to joint globe valves to pipe
- In screwed joints, the wrench should be placed on the valve end into which the pipe is inserted to prevent strain on the valve. In welded or sweated joints, the stem should be removed or backed off to avoid damage to the disc
- On water systems, the valve should be installed so that the incoming pressure is under the seat
- On steam systems, the valve should be installed so that the incoming pressure is over the seat
- On systems requiring complete drainage of the lines, the stem of valves on horizontal lines should be horizontal

Check valve

Check valves control the direction of the flow preventing the fluids from flowing into the wrong direction. The most common materials used for manufacturing check valves are brass for smaller sizes and cast iron for larger sizes.

There are two types of check valves, swing check valve and lift check valve. In the swing check valve, the flow moves straight through a tilted seat. If the flow stops, or reverses its direction, the disc closes against the seat. The swing check valve is used in pipes with moderate pressures.

The left check valve is used in pipes carrying water, air, gas, vapor, or steam and can endure higher pressures than swing check valve. The left check valves are installed on mixing valves to prevent either hot or cold water from flowing into the wrong line.

Gate valve

The operation of a gate valve is based on the screw principle. The wedge-shaped gate moves up and down at right angles to the path of flow between two rings. When seated against these rings, the gate shuts off the flow. The gate valve is installed in lines with maximum flow, for example, liquid, pump, and main lines. Gate valves should not be used for throttling as the seat may be cut by wire drawing. Gate valves may have two types of stems, rising and non-rising.

There are three types of gates that may be used on a gate valve:
- solid-wedge gate
- split-wedge gate
- double-disc gate

Solid-wedge gates are used for water, air, gas, oil, or steam lines and may be installed in any position. Split-wedge gate and double-disc gate should be installed only in vertical position.

Bypass

A bypass can be installed around the pressure-reducing and thermostatic-control valves on steam or water lines to allow the temporary use of water or steam when the valves are out of order.

The globe valve on the bypass opens when the gate valves on the line close. If a bypass around the pressure-reducing valve is used, a safety valve is required on the low-pressure side.

The safety valve should be of sufficient capacity to relieve all the fluid or steam that can pass through the bypass without over-pressuring the low-pressure side. Gauges should be installed on both high and low pressure side to observe both pressures and prevent explosion.

Pressure-reducing valve

Pressure-reducing valves are installed near the point of entrance of the water supply line to protect the water supply system from excessive pressure (over 60 pounds per square inch). Depending on the size and design of the system, pressure-reducing valves may be installed in certain zones and floors, or used to protect certain fixtures. Pressure-relief valves are mandatory in water heater, boilers, and radiators where they are installed in conjunction with safety valves to prevent explosions. Pressure-relief valves installed on water supply systems are of diaphragm type. They operate on the following principle: the low pressure on a large area overcomes a high pressure on a small area. A pressure-reducing valve consists of the following parts:
- body, with a seat facing down
- stem, that has a washer at the bottom and is attached to the diaphragm

Spring relief valve

A spring relief valve operation is based on the principle of using the tension of a coiled spring to endure the internal pressure. Spring relief valves are used on water piping, hot water tanks, pumps, boilers, air compressors, and oil lines. Spring valves are manufactured in ½- and ¾-inch sizes. The following are some of the requirements for spring relief valve installation:

- Spring relief valves should be set for the pressure 25 percent higher than the pressure on the system. This can be done by turning the adjusting screw
- A spring relief valve should have a large waterway to prevent clogging from rust scales
- All parts exposed to water should be made of brass or bronze

Lever and weight relief valve

A lever and weight relief valve operation is based on the principle of using the lever and weight to balance the pressure within a tank or a system. The weight is placed in specific locations on the lever to maintain the desired pressure. When the internal pressure exceeds the level set on the lever, the weight descends causing the seat washer to rise and relieve the pressure through the outlet. Then the weight closes the valve.

Once the position of the weight has been determined, a hole should be drilled through the weight and the lever. Then, both are fastened with a bolt. The end of the lever beyond the weight should be cut off to prevent the movement of the weight. Make sure to check the local plumbing code requirements for installation of lever and weight relief valves.

Back pressure

Back pressure is an effect opposite to siphonage. Back pressure occurs at the base of the stack where slugs of water push air before them. This causes the air pressure at the back of the stack to increase. If the horizontal drain is partially filled from other fixtures or if it receives rain water, the air pressure will increase. Back pressure can be prevented by installing a vent pipe or a relief pipe in the stack below the lowest fixture. The main vent or relief vent should be connected to the soil stack with a wye that prevents the collection of rust. This pipe should be reconnected to the soil stack above the highest fixture to let the air escape through the relief vent and prevent back pressure.

Trap

A trap is a bent tube with arms of unequal length. When the shorter end is placed under water and the air is exhausted, the pressure forces the water up the short arm, over the bend, and into the long arm. The water continues to run till the receptacle is empty. Traps may be siphoned when connected to a waste pipe. When water flows into the waste at high speed, it reduces the air pressure at the top of the stack. If the air is not replaced, the trap will be siphoned. Water and sewage flowing into the soil stack can form slugs that cause a greater air pressure at the bottom and the lower air pressure at the top of the stack. The greater air pressure is called plenum; the lower pressure – partial vacuum. The water siphons through the trap seals allowing drainage gases into the building.

A P-trap installation

A P-trap that consists of a crown, a crown weir, a bottom dip, a top dip, and a seal should be installed as close to the fixture as possible to prevent fouling of the inlet side of the trap. P-traps can be used for most of the plumbing fixtures except water closet that are manufactured with an integral S-trap. When P-traps are installed with wall-hung fixtures, the vertical inlet leg of the trap must not exceed 24 inches between the fixture outlet and the trap weir. When P-traps are installed with floor-set fixtures, the trap should be concealed between the building joints. Sizes of P-traps depend on the type of the plumbing fixture and vary from 1¼ to 6 inches. Most of the fixtures require 1½-inch size trap, for example, bathtubs, domestic dishwashers, and sinks. Pedestal urinals, stalls, and floor drains require minimum 3-inch size traps and larger.

Interceptors

Interceptor, or separator, is a term that refers to a part of a drainage system that catches objectionable waste, such as oil, grease, sand, plaster, lint, hair, and glass and separates it from the other liquid waste.

The purpose of the use of an interceptor is to prevent the harmful objectionable waste from entering the building's drainage system. Oil and sand interceptors should be installed with the approval of the Plumbing Official or other administrative authority only in the areas where it is necessary to protect the building drainage system from harmful oils and sands. According to the Code, interceptors should be designed so that they do not become air-bound.

Grease interceptors installation requirements

The following are the main requirements for installation of grease interceptors:
- Grease interceptors should be installed on waste lines conveying waste from sinks, drains, and other fixtures in restaurants, hotel kitchens, bars, clubs, factory cafeterias, hospital kitchens, and any other commercial or industrial establishment that introduces large amounts of grease into the sewage system
- Grease interceptors should not allow siphonage or become air bound

- Grease interceptors should be placed as near to the fixture as possible while at the same time being placed outside of the building. Interceptors may be installed on the floor, or flush with the floor
- Grease interceptors should be protected from freezing

The following are the requirements for installation of grease interceptors in garages, beauty shops, and similar establishments:
- Grease interceptors should be installed in establishments where motor vehicles are serviced and where excessive sediment and hair are discharged into the drainage
- In garages and other vehicle servicing establishments, the size of the fixture catch basin should be allowed to hold oil, sand, and dirt discharging into the interceptor during any 10-hour period. The minimum size of the basin should be 24 inches wide and 24 inches deep
- Grease interceptors in beauty shops should have a removable perforated copper basket with a sloping bottom that has a blowout neck standard

Grease interceptors efficiency

The following are the requirements for efficiency of grease interceptors:
- Grease interceptors should be tested and be in accordance with these standards: ASME A112.14.3, ASME A112.14.4, and PDI-G101
- Grease interceptors should have efficiency ratings that comply with the local standards and practices
- The flow rate of a grease interceptor should not exceed its rated capacity
- A mean efficiency rate should be at least 90 percent
- Accumulated grease should be periodically removed in order to maintain interceptors in efficient operating condition and prevent grease and foreign material from entering into the drainage system
- If the amount of grease discharged into the fixture drain increases, traps should be installed

Running trap installation

A running trap is a U-shaped water-sealed device that is usually used to provide trap seal protection on area drains, rainwater leaders, and downspouts. In the areas where the local plumbing code requires trapping of the building drain, a running trap can be installed on the house traps. A running trap can be made of the following types of piping:
- DWV copper
- Hub-style soil pipe
- Cast iron drainage

Pipes made of other materials are also available. The size of the running trap depends on the plumbing fixture it is used with and ranges from 1¼ to 4 inches. As all traps are subject to stoppage, a running trap must be provided with a clean out that should be easily accessible and disassembled.

Sand trap

Sand traps are required for installation in garages because the drainage containing oils, grease, and sand may clog a regular drain or sewer. A sand trap consists of an 18-inches deep catch basin with a cover. The top of the trap should be level with the floor. The outlet should be taken from an elbow turned down below the water surface.

Gas and oil do not enter the drain because they are lighter and float on the surface. The sand settles at the bottom. It is important to clean the sand traps regularly to prevent clogging. In large garages, two-compartment sand traps are used to avoid the entrance of gas and oil into the sewer.

Trap seal primer

The main purpose of a trap seal primer is to maintain a trap seal installed on cold water supply lines to kitchen sinks, toilets, drinking fountains, and similar fixtures with connections to traps on rarely used waste lines. The water seal might evaporate from the trap and allow sewage gas and vermin to penetrate into the building through the sewer lines. A trap seal primer valve ensures that there is a permanent seal of fresh water in the drain trap. The principle of operation of a trap seal primer is as follows:
- The fixture draws water through the supply line
- The flow activates the valve that measures an exact amount of water from the supply line into the trap line
- The vacuum breaker installed on the primer valve prevents siphonage

Shower drain installation

Pans
The following are the International and State Plumbing Codes requirements for drain installation in lead and copper shower pans:
- The drain installed should not allow the water accumulating in the pan to leak around the drain and down the exterior of the pipe
- The drain must have a flange located beneath the pan material that should be cut to allow water into the drain. Another part of the drain should be placed over the pan material and secured to the bottom flange. This will create a compression between the top and the bottom flange and ensure that the seal is watertight
- The strainer should be screwed into the bottom flange housing. It can be screwed up or down, depending on the level of the finished shower pan

<u>Requirements</u>
The following are the International and State Plumbing Codes requirements for shower pan installation:
- The surface of the floor where the shower pan is going to be installed should be smooth and able to support the weight of the shower
- Shower pans must be made from a waterproof material, such as flexible vinyl
- The edges of the pan material must extend at least 2 inches above the height of the threshold. Some codes require 3-inch extension
- The threshold must be at least 2 inches, or higher, but no higher than 9 inches
- In handicap facilities, the threshold may not be required
- The pan material must be secured to the stud wall
- The lining must not be nailed or perforated anywhere less than 1 inch above the finished threshold

ADA requirements

<u>Toilet stalls</u>
The following are the ADA requirements for toilet stalls:
- Toilet stalls should be located on an accessible route
- The minimum width of the stall should be 60". The centerline of the water closet should be 18" from the side wall
- The front partition and at least one side partition should provide a toe clearance of at least 9" above the floor. If the depth of the stall is greater than 60", the toe clearance is not required
- Toilet stall doors, including door hardware, should be accessible. If toilet stall approach is from the latch side of the stall door, clearance between the door side of the stall and any obstruction may be reduced to 42"
- The side grab bar should be 40-42" in length, beginning 12" from the rear wall, 33-36" above the floor. Grab bars should have a gripping surface and should not obstruct the required clear floor area

<u>Water closet</u>
The following are the ADA requirements for water closets:
- Depending on the type of the transfer to the water closet, the minimum clear floor space at the water closet should be 48" by 66", 48" by 56", or 60" by 56". Clear floor space should allow either a left-handed or right-handed approach
- The height should be 17" to 19" to the top of the toilet seat. Seats should not be sprung to return to a lifted position
- Grab bars behind the water closet should be 36"
- Flush controls should be hand operated or automatic and mounted on the wide side of toilet areas no more than 44" above the floor
- Toilet paper dispensers should be within reach. Controlled delivery dispensers are not permitted

Urinal

The following are the ADA requirements for urinals:

- Urinals should be stall-type or wall-hung with an elongated rim at a maximum of 17" above the finish floor
- A clear floor space 30" by 48" should be provided in front of urinals to allow forward approach. This clear space should adjoin or overlap an accessible route and should be accessible by wheelchair. Urinal shields that do not extend beyond the front edge of the urinal rim should be provided. There should be 29" clearance between them
- Flush controls should be hand operated or automatic and mounted on the wide side of toilet areas no more than 44" above the finish floor

Sink

The following are the ADA requirements for sinks:

- Sinks should be mounted with the counter or rim no higher than 34" above the floor
- Knee clearance should be at least 27" high, 30" wide, and 19" deep
- The depth should be 6½"
- A clear floor space should be at least 30" by 48" and should allow a forward approach. The clear floor space should be on an accessible route and extend to 19" underneath the sink
- Hot water and drainpipes exposed under sinks should be insulated and protected against contact. No sharp or abrasive surfaces should be under sinks
- Faucets should be operable with one hand and should not require tight grasping, pinching, or twisting of the wrist. The force required to activate controls should not exceed 5 lbf (22.2 N). Lever-operated, push-type, touch-type, or electronically controlled faucet mechanisms are acceptable

Drinking fountain

The following are the ADA requirements for drinking fountains:

- Spouts should be no higher than 36" from the floor or ground surfaces to the spout outlet
- The spouts of drinking fountains should be at the front of the unit and direct the water flow in parallel or nearly parallel to the front of the unit
- The spout should provide a flow of water at least 4" high to allow the insertion of a cup or glass under the flow of water
- Controls should be operable with one hand and should not require tight grasping, pinching, or twisting of the wrist. Unit controls shall be front mounted or side mounted near the front edge
- A clear knee space between the bottom of the apron and the floor should be at least 27" high, 30" wide, and 17-19" deep for wall- and post-mounted cantilevered units. These units should have a clear floor space 30" by 48"

Lavatory

The following are the ADA requirements for lavatories:

- Lavatories should be mounted with the rim or counter surface no higher than 34" above the floor. A clearance of at least 29" above the floor to the bottom of the apron should be provided
- A clear floor space 30" by 48" should be provided in front of a lavatory to allow forward approach. The clear floor space should adjoin or overlap an accessible route and should extend to 19" underneath the lavatory
- Hot water and drainpipes exposed under sinks should be insulated and protected against contact. No sharp or abrasive surfaces should be under lavatories
- Faucets should be operable with one hand and should not require tight grasping, pinching, or twisting of the wrist. The force required to activate controls should not exceed 5 lbf (22.2 N). If self-closing valves are used the faucet should remain open for at least 10 seconds

Shower stall

The following are the ADA requirements for shower stalls:

- The shower stall should be at least 36" by 36"
- A seat should be provided in shower stalls. The seat should be mounted 17" to 19" from the bathroom floor and extend the full depth of the stall. The structural strength of seats and their attachments should be no less than 250 lbf
- Grab bars should not rotate within their fittings
- Controls, faucets, and the shower unit should be mounted on the sidewall opposite the seat
- A shower spray unit with a hose at least 60" long that is used both as a fixed showerhead and as a hand-held shower should be provided
- Curbs should be no higher than ½"
- Shower enclosures should not obstruct controls or obstruct transfer from wheelchairs onto shower seats

Bathtub

The following are the ADA requirements for bathtubs:

- Depending on the design of the bathtub, the clear floor space in front of bathtubs should be 30" by 60", 48" by 60", or 30" by 75". The seat width must be 15 inches and extend to the full width of the bathtub
- An in-tub seat or a seat at the head end of the tub should be provided. The structural strength of seats and their attachments should be no less than 250 lbf. Seats should be secured and not slip
- Grab bars should not rotate within their fittings
- Faucets should be operable with one hand and should not require tight grasping, pinching, or twisting of the wrist
- A shower spray unit with a hose at least 60" long that is used both as a fixed showerhead and as a hand-held shower should be provided

Traps and cleanouts installation

The following are the requirements for installation of traps and cleanouts on indirect waste piping:

- When the length of the piping exceeds 2 feet horizontally or 4 feet of total developed length, traps should be installed
- Venting of the traps is not required because indirect waste pipes do not add offensive odors to a receiving fixture, such as floor sink, floor drain, open pipe, or pipe hub
- The pipe should be properly supported to avoid accumulation of slime and development of stoppages due to low velocity of the wastewater in the pipe
- Accessible cleanouts should be installed so that the pipe could be cleaned and flushed

Indirect waste systems requirements

The following are the requirements for indirect waste systems:

- Indirect waste systems should be installed on refrigerators; ice boxes; steam tables and other devices where food is processed and stored; stills, sterilizers, water coolers; commercial dishwashing machines; overflows, drains, and relief vents of water supply systems
- Wastes from the appliances and fixtures described above should discharge into the structural drainage system through indirect piping that should be provided with an air gap of a minimum double-size of the diameter of the drain served
- Laundry sinks and similar fixtures may be provided with an extended indirect waste pipe that has an air gap in the drain connection on the inlet side of the trap installed in the fixture

Septic tank

A septic tank is a watertight container that receives waste from a drainage system through an inlet tee. The purpose of the septic tank is to separate liquid waste from solid that makes up approximately ¾ of a pound in every 100 gallons of water. Solids accumulate at the bottom of the tank while liquids and grease rise to the top. In the absence of air, anaerobic bacteria decompose the solids converting them into harmless gases and liquids. The gases escape into the atmosphere through the drainage vent pipes. A clear liquid, called effluent, is the result of the work of bacteria. The effluent is forced into the drain field though the outlet tee. At disposal fields aerobic bacteria decompose the effluent further. The Code requires a septic tank to be large enough to hold as much sewage water as is expected to flow into the tank in 24 hours.

<u>Capacity</u>
The number of persons using the facilities served by the septic tank is the main factor in determining the size of the tank. The minimum capacity is 500 gallons. The following table illustrates minimum required sizes of septic tanks servicing one- and two-family residences:

Bedrooms	Typical Plumbing Fixtures	With garbage disposal system, dishwasher, and washing machine
2 or less	500 gallons	750 gallons
3	700 gallons	975 gallons
4	850 gallons	1200 gallons
5	950 gallons	1400 gallons
6	1150 gallons	1650 gallons

<u>Requirements</u>
The following are the requirements for septic tanks:
- Septic tanks should be watertight and have individual structures made of monolithic concrete, welded steel, or other approved material
- The inside diameter of cylindrical tanks should be no less than 48 inches
- The inside width of rectangular tanks should be no less than 36 inches. The longest dimension should be parallel to the flow direction
- The walls of concrete septic tanks should be no less than 2 inches
- Tanks should be marked to indicate liquid capacity and manufacturer's name and address
- Concrete tanks should endure usage pressures
- Steel tanks should be made of hot-rolled commercial steel of a minimum thickness of 14-gage

<u>Sludge disposal</u>
Usually, the local plumbing code administrative authorities require that septic tank sludge should be disposed at a predetermined site or into a public sewer. If these facilities are not available, the following procedures should be used:
- Septic tank sludge should be buried under 42 inches of earth. The burial should be located at least 55 feet from a well or water source and at least 550 feet from dwellings. There should be at least 42 inches of soil between the buried septic tank sludge and the high ground water level
- Septic tank sludge may be spread on land which is not used for crops cultivation or livestock pasture. The spread should be located at least 1,200 feet from any water source or dwelling

Installation requirements

The following are the requirements for septic tank installation:

- Septic tanks should not be installed within the interior foundation walls of a building
- New buildings should not be constructed within 5 feet of an existing septic tank
- A bedding base is required for placement of septic tank. The bedding base can be made of gravel, sand, granite, or other approved material that should be well tamped and 3 inches thick
- Materials used for bedding bases should not be above the standard size 14 in crushed material
- Concrete tanks should be backfilled with soil materials that should be well tamped and be no more than 2 inches in size
- Steel tanks should be backfilled with gravel, sand, crushed limestone, or granite not be above the standard size 14 in crushed material

Location requirements

The following are the requirements for septic tank installation location:

- The distance from the building should be no less than 5 feet
- The distance from the soil absorption system to the building should be no less than 25 feet, to any water source – no less than 55 feet, to any lot line – no less than 8 feet, to water service – no less than 30 feet

Residential trailer

Fixtures installation

The following are the general recommendations for installation of fixtures in residential trailers:

- Fixtures should be made of nonabsorbent materials
- Fixture installation should allow the fixtures to endure road shock and vibration
- Fixture traps should have a water seal no less than 2 inches thick
- Water closets should be made of durable materials
- Installation of water closets should not allow the fixtures to spill the trap seal contents during travel
- Water closets should be protected against flushing unless connected to an approved sewage disposal source
- Connection between a trailer drainage and a trailer park sewage system should be made with a watertight flexible or semi-rigid connector
- Water closets should have sufficient water supply to clean the interior of the unit and be equipped with a vacuum break to prevent water supply contamination

- 94 -

<u>Drainage and vent systems</u>
The following are the requirements for drainage and vent systems used in residential trailers:
- Residential trailer toilet compartments should have proper light and venting facilities
- Piping and conduit openings through floors and walls should be permanently sealed to prevent entrance of rodents and vermin
- Horizontal drainage piping should be installed in alignment at a slope of no less than ⅛ inch per foot
- Contraction and extension should be considered while aligning piping
- Fixtures should be secured and protected against freezing
- Exterior piping, fixtures, and appliances may be installed as long as they don't interfere with the operation of doors and windows
- Only new materials must be used for water supply systems

Continuous vents

A continuous vent is an extension of the vertical waste pipe that leads past the fixture, back into a branch vent, vent stack, or stack vent. The most common materials used for manufacturing continuous vents are lead, copper, brass, galvanized, wrought iron, cast iron, or plastic. When a continuous vent is installed, the following rules should be observed:
- The size of the vent pipe should be no less than half of the waste pipe but never less that 1¼ inch
- The waste pipe tee should not be installed below the trap seal to prevent siphonage
- The continuous vent should be installed as close to the fixture as possible to prevent clogging of the vent

New home water treatments

Such new water treatment technologies as water treatment plant provide for a cleaner water emission and are able to reduce the typical household wastewater to clear odorless liquids in 24 hours. The method of operation of these water treatment plants is as follows:
- The primary treatment compartment receives the wastewater and holds it until the solid matter settles to the sludge layer at the bottom of the tank
- Anaerobic bacteria break down organic solids
- Inorganic solids, such as grit, are settled out and held back
- The aeration compartment receives the finely divided material and mixes it with activated sludge
- Aerobic bacteria break the material further and convert it into the odorless mixture of liquids and gases
- The clarifying compartment receives the material and sends it back to the aeration compartment if there is any floating material

- 95 -

Storm water drainage system

The following are the components of the storm water drainage system:
- Storm sewer – used for conveying groundwater, rainwater, surface water, or other non-pollution wastes
- Building storm sewer – a sewer that conveys only storm water but no sewage
- Building storm drain – a drain that conveys only storm water but no sewage
- Roof drain – a drain installed on the roof that receives water accumulating on the surface of the roof and conveys it into a rainwater leader, conductor, or downspout
- Storm drain cleanout – a cleanout placed in the rainwater leader, conductor, or downspout to afford access to the pipe for the purposes of cleaning from foreign material

Neutralizing tanks installation

A neutralizing tank is installed in medical and other facilities handling corrosive liquids, spent acids, and other materials that can damage the regular drainage system and cause health risks. Before entering the drain, the waste should be treated. A neutralizing tank is one of the systems that serve the treatment purpose. The tank should have capacity large enough to provide water or other chemical agent to dilute and neutralize toxic waste before it is discharged into the drainage system. Selection of the neutralizing tank capacity depends on the amount and composition of the corrosive waste discharged into it. The tank should be provided with an access cover and have an inlet and an outlet that should be located below the inlet level.

Strainer requirements

The following are the requirements for strainers according to the Uniform Plumbing Code:
- Roof drains should be equipped with strainers that should extend no less than 4 inches above the surface of the roof adjoining the drain
- The area of a strainer should equal to no less than 1½ of the pipe it is attached to
- Roof drain strainers on sun decks, parking decks, and other occupied areas should be of an approved flat-surface type and level with the deck
- The area of a roof drain strainer on sun decks, parking decks, and other occupied areas should be twice as much as that of the pipe it is attached to roof drains running through the interior should be made watertight with the help of suitable flashing material

Rainwater piping requirements

The following are the requirements for rainwater piping according to the Uniform Plumbing Code:

- Rainwater pipes should not be used in place of soil, waste, or vent pipes
- Soil, waste, or vent pipes should not be used in place of rainwater pipes
- Rainwater piping must be protected in the areas where they can be potentially damaged
- Any roof drains, overflow drains, and rainwater piping installed within the construction of the building should be tested for conformity with the regulations of the UPC
- Rainwater piping installed within the building should be made of cast iron, galvanized steel, wrought iron, brass, copper, lead, or other approved materials
- Rainwater piping installed on the exterior should be made of no less than 26 ga galvanized sheet metal

Horizontal rainwater piping sizing

The horizontal rainwater piping is sized according to the UPC Rainwater Sizing Tables. The sizing is based on maximum roof areas to be drained. The following example illustrates the calculation:

- Roof Area = 5900sq.ft.
- Max. Rainfall/hr. = 5 inches
- Pipe Slope = ¼ inch

In the UPC Rainwater Sizing Table find the roof area (5900sq.ft.) under the 5" rainfall column. Read 6" as the pipe size.

The semi-circular roof gutters are sized according to the UPC Rainwater Sizing Tables. The sizing is based on maximum roof areas to be drained. The following example illustrates the calculation:

- Roof Area = 2000sq.ft.
- Max. Rainfall/hr. = 4 inches
- Gutter Slope = ⅛ inch

In the UPC Rainwater Sizing Table find the roof area (2000sq.ft.) under the 4" rainfall column. Read 7" as the gutter diameter in the left hand column.

Vertical rainwater piping sizing

Vertical rainwater piping should be sized according to the UPC Rainwater Sizing Tables and is based on maximum inches of rainfall per hour falling on particular roof area. To calculate the total area of the walls and roof, add the following to the roof area:

- For one wall add 50 percent of the wall area
- For two adjacent walls add 35 percent of the wall areas sum
- For two same height opposite walls add zero
- For two deferent height opposite walls add 50 percent of the wall area above top of the lower wall
- For three walls calculate the area of the two walls as in (2) or (3), then add 50 percent of the inner wall area below the lowest wall
- For four walls on all sides add areas above the top of the lowest wall calculated according to (1), (2), (4), and (5)

Perform System Testing

Air test

The following is the procedure for the air test of water supply and distribution piping:

- An air compressor must be attached to any suitable opening
- All other openings must be closed with a pipe cap, a pipe plug, or the system control valve all of which usually hold pressure better than test plugs
- Air should be forced into the system until it reaches a pressure that is 1½ times greater than the working pressure, or 150 pounds per square inch. The greater figure should be selected as the suitable pressure
- This pressure should be maintained for a period from 12 to 24 hours without adding more compressed air

Finished plumbing test steps

The following are the steps for finished plumbing test:

- Fill all fixture and floor drain traps with water
- Plug all stack venting openings on the roof
- Plug the building drain at the front main cleanout
- Insert the manometer hose through the trap seal of a water closet bowl and blow any water out of the hose
- Fill the manometer with water to the 0 mark on the ruler
- Attach the hose to the manometer and set the manometer to rest on the open closet seal
- Blow air into the system using another hose inserted through the trap of a water closet. Blow air till it reaches and maintains a 1-inch differential of pressure

If the manometer does not hold the pressure of 1 inch of water column, there are leaks in the system

Hydrostatic test

The hydrostatic test of water supply and distribution piping consists of filling the pipe with water and then applying additional water pressure. Water main and water service piping are usually tested with the hydrostatic method. The following is the procedure for the hydrostatic test:

- All openings should be sealed with pipe caps, pipe plugs, or test plugs
- The pipe should be filled completely with potable water
- A hydrostatic test pump should be used to force additional water into the pipe. The pump should be equipped with check valves for suction and

discharge ends to prevent the water pressure from forcing after back through the pump
- The additional water should not be excessive. For example, a 30-gallon piping system requires only extra 150 ml of water that increases the pressure to 50 pounds per square inch
- All air pockets should be eliminated for successful testing

Peppermint test

The following is the procedure for the peppermint test:
- Fill all fixture and floor drain traps with water
- Plug the building drain at the front main cleanout
- Pour two ounces of peppermint oil into each stack and the building drain
- Plug the stack and the building drain openings
- Insert the manometer hose through the trap seal of a water closet bowl and blow any water out of the hose
- Fill the manometer with water to the 0 mark on the ruler
- Attach the hose to the manometer and set the manometer to rest on the open closet seal
- Blow air into the system using a hose inserted through the trap of a water closet. Blow air till it reaches and maintains a 1-inch differential of pressure

If the manometer does not hold the pressure, there are leaks in the system that can be found using the sense of smell.

Sanitary drainage and vent system

The plumbing specialist should test the sanitary drainage systems when the sanitary drainage and vent piping installation has been completed. After the test, the systems should be inspected by an authorized administrative officer.

The following are the methods used to test sanitary drainage and vent systems:
- Water test – can be applied to the entire system or its sections; consists of running pressured water through the system and noticing any leaks
- Smoke test – a smoke machine is used to produce an odorous smoke that is pumped through the system to locate any leaks
- Air test – an air compressor is used to fill the system with pressured air, a soap solution is used to locate leaks
- Peppermint test – a peppermint oil solution is used to fill the system; leaks are located by the sense of smell

Air test

The following items are necessary for conducting the air test of sanitary drainage and vent piping:

- An air compressor
- Caps and plugs to close inlets and outlets of the system
- A test gauge assembly
- A soap solution to check for leaks

To facilitate the air test of sanitary drainage and vent piping, an air compressor must be attached to any suitable opening. All other openings must be closed with a pipe cap, a pipe plug, or a testing plug. Air should be forced into the system until it reaches a pressure of 5 pounds per square inch of the section being tested. This pressure should be maintained for 15 minutes without adding more compressed air.

Smoke test

The following is the procedure for the smoke test of sanitary and drainage systems:

- The fixture traps should be filled with water
- The building drain at the front main cleanout should be plugged
- Odorous thick smoke should be produced with the help of a smoke machine and introduced into the system
- When smoke appears at the opening on the roof, the opening should be sealed
- A pressure of 1 inch of water column should be built up and maintained within the system for 15 minutes
- The leaks in the system are detected by using the smell sense, so the plumber should avoid any of the smoke odors on the clothing

Pipe caps and plugs

Pipe caps and plugs are used to hold the compressed air in the sanitary drainage and vent piping system during testing. Appropriate caps and plugs must be selected for the testing. For example, for closing threaded iron piping, iron pipe caps and plugs should be used; for closing copper piping, soldered caps should be used; for closing plastic pipe openings, plastic pipe caps and plugs should be used.

Sometimes, pipe caps and plugs cannot be used to close openings in the sanitary drainage and vent piping. In these cases, a mechanical or inflatable rubber test plugs should be used. A mechanical test plug consists of a capped stem of ½-inch pipe equipped with a running thread. Inflatable rubber plugs are inverted into the piping opening and inflated with a tire pump.

Air compressor selection

The selection of a suitable air compressor depends on the size of the plumbing system. For single-family dwellings, the following devices can be used:
- A hand-pump
- A small air-compressor

Many manufacturers offer three types of small air compressors: Invector, Direct-Drive and Belt-Drive. Invector air compressors are equipped with a universal motor that uses an air-cooling system. Direct-Drive compressors feature standard induction motors. Belt-Drive air compressors require oil for operation.

For testing of larger installations, the following devices should be used:
- Gasoline motor driven air compressor
- An air hose to pipe the compressed air from the compressor to the test opening

Motor air compressors can be equipped with the following: 10" Pneumatic Inlet Valve, One 3" Service Connection, Single Point Lifting Bail, Tri-Axle Running Gear, Low Fuel Shutdown, Automatic Shut Down and Protection System, Direct Drive, Adjustable 3" Lunette Eye.

Testing gas pipe pressure requirements

The following are the National Fuel Gas Code requirements for methods of testing gas pipe pressure:
- Using a manometer or another pressure measuring device designed and calibrated to read, record, and indicate a pressure loss resulting from leakage
- Isolating the source of pressure before testing
- Using the test pressure that is no less than 1½ of the operating pressure but does not go below 3 psi regardless of the operating pressure. If the test pressure reaches beyond 125 psi, it still should not exceed a value that produces a hoop stress greater that 50 percent of the minimum yield stress of the pipe
- Keeping the test duration under 30 minutes for each 500 cubic feet of pipe volume
- Keeping the test duration under 10 minutes if the pipe volume does not exceed 10 cubic feet

Sanitary drainage water test

The following are the water test procedure for sanitary drainage systems:
- Plug all openings except the highest openings
- Fill the system with water through the highest opening or through the valve and hose connection placed in one of the fixture drain openings
- In tall buildings, the sanitary drainage systems are checked in sections in order to avoid high pressure of the water column. To determine how many sections can be tested at a time, it is necessary to take into consideration that every 1 foot of head of water equals 0.434 psi
- Check for and locate water leaks
- Remove the water from the system by releasing the plug, or using a water removal plate when dealing with large quantities of water
- Repair the leaks

Sterilizing potable water systems

The following is the procedure for sterilizing potable water systems:
- Flush the pipe system with clean potable water until dirty water does not appear at the outlets
- Fill the system or a part of it with a water-chlorine solution containing at least 50 mg/l of chlorine
- Valve off the system and allow to stand for 24 hours. If a 200 mg/l water-chlorine solution is used, allow the system to stand only for 3 hours
- Flush the system with clean potable water until all the chlorine is gone and is not coming from the system any longer

Repeat the procedure if a bacteriological analysis conducted by the authority shows that there is still contamination in the system.

Backflow test certification

The backflow test should be certified in order to be legitimate. A trained certified plumbing specialist should perform the test and ascertain its compliance with the national and local plumbing codes. The certification is conducted through an exam by a local backflow prevention program implemented under the supervision of a State Environmental Protection Department. In addition, the American Backflow Prevention Association, the American Society of Sanitary Engineering, and the University of Florida/TREEO Center have been approved as official certification programs. The certified tester must check the following:
- Check valve 1 and 2
- Pressure-relief valve
- Pressure-vacuum break

The meter number should be identified. Then, the certified tester fills a form to ascertain the quality of the backflow prevention system.

Gas supply lines test preparation

The following are the methods of gas supply lines test preparation:
- Exposing all pipe joints, including welds
- Providing expansion joints for additional thrust load
- Disconnecting all appliances and equipment, or isolating them by using caps, blind flanges, or blanks
- Disconnecting and capping the outlets of all appliances and equipment that are designed to operate at pressures lower than the one required for the test
- Closing the shutoff valve of each individual appliance which is designed to operate at pressures equal or higher than the one required for the test
- Clearing the interior of the pipe of all foreign material
- Taking the necessary safety measures to protect employees and general public from potential dangers

Potable water certification

After sterilizing a potable water system, a sample of water should be obtained and sent for analysis to a laboratory. The laboratory should be certified by the State Department of Health. In the laboratory, water is tested for coliform bacteria, nitrate, lead, and arsenic. To be acceptable, the water's test results must be less than two years old and comply with the following minimum requirements:
- Total Coliform – None present
- Nitrate – Less than or equal to 10 mg/L
- Lead – Less than or equal to 0.05 mg/L
- Arsenic – Less than or equal to 0.05 mg/L

The test results are used to obtain certification from the State Department of Health.

Perform maintenance and repair

Soapy water method

To test gas supply lines for leaks, a soapy water solution can be used. The following are the steps for conducting the soapy water solution leakage test:
- Mix any common household soap detergent, such as dishwasher detergent, washing machine detergent, or just liquid hand soap with water in a small container
- Brush the soapy water solution on all the joints of the gas supply piping
- Observe each joint of the piping for bubbles. The soapy mixture will produce vigorous bubbles if there are leaks in the joints of the piping

Detecting defects and leaks

The following are the National Fuel Gas Code requirements for detecting defects and leaks in gas piping:
- Gas piping system should endure the test pressure without indication of leaks or defects
- If pressure-measuring devices indicate reduction of pressure, it should be interpreted as evidence of leaks of defects
- An approved gas detector, a non-corrosive leak detection fluid, or other approved leak detection method must be used to locate the leakage
- Any methods that provide a source of ignition such as matches, candles, and open flames, must not be used
- The piping section where the leakage or defect has been detected must be repaired or replaced, and retested

Checking gas burner's input

The following are the Standard Gas Code requirements for checking gas burner's input without using a meter:
- Gauge the size of the burner orifice and calculate flow rate at sea level using Table F1 A or B (Annex F Flow of Gas Through Fixed Orifices of National Fuel Gas Code, 2006 edition)
- When the utility gas gravity is other than 0.60, multiply the flow rate by the number found in Table F1 C under the specific utility gas gravity
- When the altitude is above 2000 feet, select the orifice size at sea level from Table F1 D, then determine flow rate using Table F1 A or B. Check the burner input according to the manufacturer's input rate

Gas appliance pressure regulators

The following are the National Fuel Gas Code requirements for venting of gas appliance pressure regulators:
- A vent piping leading outdoors should be provided for appliance pressure regulators that require access to the atmosphere
- A vent piping leading into the combustion chamber should be supplied with appliance pressure regulators that have the vent piping as an integral part
- Only listed appliance pressure regulators should employ vent-limiting means
- A vent piping leading outdoors should be protected from penetration of water and foreign objects, such as insects and leaves
- An appliance regulator should never be vented to the appliance flue or exhaust system
- For vents leading into the combustion chamber, the location of the vent should facilitate the quick ignition of the escaping gas
- A vent line and a bleed line should not be connected to a common manifold leading into the combustion chamber

Electric water heater element leakage

The following might be the causes of leakage at the elements:
- Defective elements that leak at terminals or through flange
- Loose elements
- Loose gaskets

Leakage repair
The following are the methods of repairing the leaking elements:
- To stop the leakage at the defective elements, replace the element. Before removing the element, it is necessary to shut off electricity and drain the tank
- To repair a loose screw-in gasket, tighten with a 1½ socket wrench or Part No.21163 wrench. If the leak persists, remove the element, discard the gasket, clean the thread areas, and apply non-hardening Permatex No.2. Install a new gasket, screw, and tighten the element
- To repair a loose flange gasket, tighten with a wrench. If the leak persists, remove the element, discard the gasket, clean the gasket seating areas, and reinstall the element with a new gasket

Checking the electric water heater

Water expansion during the heating cycle as well as some foreign material on the seat of the valve may cause relief valve and drain valve leakage. In addition, released water may be mistaken for leakage if the valve is not piped to an open drain. To check where the threaded portion enters the tank, insert some absorbent material, such as a Q-tip, between the jacket opening and the valve. Swab the spud area. If there is an indication of a leak, remove the valve and repair with pipe joint compound. Before removing either the relief valve or the drain valve, it is necessary to shut off electricity and drain the tank.

Repairing stoppage of building sewer

The following are the steps to repair stoppage of the building sewer:
- Access the sewer through the cleanout located near the basement wall or by excavating and breaking through the pipe
- Use these tools to free the drain from roots, grease, branches, and hard objects: a) plunger, b) root cutter, c) screw, d) claw, e) spear point, f) hinged hoe scraper, g) sand scoop, and h) sewer brush
- Use copper sulfate to kill tree roots and prevent drain stoppage in one of the following ways: a) pour the chemical into the drain, avoid flushing for a while; b) pack copper sulfate around the outside of pipe joints at the time of laying the sewer; c) place copper rings in pipe joints

Repair stoppage of lavatories

The following are the steps to repair stoppage of lavatories and water closets:
- Cover the overflow to concentrate the pressure in the drain
- To dislodge the stoppage and force it down the drain, fill the fixture with water and surge vigorously with the pneumatic plunger
- If the stoppage in the lavatory does not clear up, drain the contents of the fixture through the cleanout plug in the bottom of the drain
- Remove the strain and clean the drainpipe with a flexible wire or brush
- If the stoppage in the water closet does not clear up, use a flexible wire or brush
- If the stoppage does not clear up after this, the branch or main drainpipes might be blocked. Clean these pipes through their cleanouts with a flexible wire

Removing obstructions in drains

Specific conditions of the system should guide the plumber in locating obstructions in the drain. If the fixtures do not drain properly and water overflows at the fresh air inlet, the obstruction might be in the main house trap. The obstruction can be removed by inserting a grappler through the fresh air inlet or by using a plunger. If the obstruction is located between the trap and the sewer, it can be eliminated by removing the cleanout plug on the sewer side and inserting a coil wire. This has to be done carefully to avoid the increase of back pressure and flooding the area with sewage water. It is possible to locate obstructions by tapping the pipe with the hammer. The pipe will produce a solid sound when the hammer taps on the obstruction.

Cleaning water service pipes

The following are the steps to clean water service pipes:
- Stop the flow in the pipe by inserting a valve or by freezing
- Attach a scraper or a cutting tool to a flexible metal tape or spring that is twisted by hand or by power as it passes through the pipe
- Push or pull the scrape or the cutting tool through the pipe
- Alternatively, use a pneumatic tool, available for clearing clogged pipes, which delivers a slug of compressed air at a pressure of 60 to 80 psi
- In severe stoppage cases, release a mixture of air and water at a pressure of over 1,000 psi

Repairing ball-type shower faucets

The following are the steps to repair ball-type shower and bath faucets:
- Shut down water supply valves and drain the lines by turning on the faucet handles
- If dealing with a lever-type handle, locate the set screw, and then loosen and pry the handle off the stem
- If dealing with a round-type handle, remove the decorative cap, locate the set screw, and then loosen and pry the handle off the stem
- Cover the cap with the tape, unscrew the handle counterclockwise
- Lift out the seats and springs with a pencil or a sharp tool
- Check the parts for wear and damage
- Replace the damaged or worn parts
- Reassemble the faucet making sure that the slot in the ball slips into the pin in the valve body and the lug on the cam assembly slides into the slot on the valve body

Maintenance of backwater valves

Provided the backwater valve was installed in accordance with the BOCA National Building Code Section P-1003.3 "Location of backwater valves", maintenance and cleaning of the valve should not present any serious problems. The Code states "Backwater valves shall be installed so their working parts will be readily accessible for service and repairs."

All backwater valves should be equipped with cleanouts through which the access to the valve could be gained. To ensure that the valve operates properly, it is necessary to check the valve annually for free movement of all moving parts and remove any material that may prevent free movement. If the valve is clogged, clean away any debris that may have accumulated inside the trap.

Leaks in water closet flush valves

The following are the possible causes for leaks in water closet flush valves:
- The float on the float valve is submerged more than it should be and water is escaping through the overflow
- The flush valve is not seated and water runs into the toilet bowl

If the float on the float valve is water-clogged, replace it with a new float. If the inlet valve is leaking, replace the seat or repack. If neither the float nor valve is the problem, bend the bar holding the float so that the float can rise sooner and close the valve tightly.

If the flush valve is not seated, adjust or replace the rubber ball, the guide, or the valve itself.

Repairing cartridge-type shower valves

The following are the steps to repair cartridge-type shower and bath valves:
- Shut down water supply valves and drain the lines by turning on the faucet handles
- Pry off the decorative cap and remove the handle screws
- Remove the handle and escutcheon
- Use pliers to remove the retainer clip
- Pull the cartridge off the faucet body
- Note the position of the ears on the cartridge to place them back correctly while reassembling
- Check the O-rings and cartridge for wear and damage
- Replace the damaged or worn parts where necessary
- Reassemble the valve by using the reverse procedure
- Turn on water and check for leaks

Locating leaks with smoke test

To locate leaks in sanitary drainage and vent systems using a smoke test follow these steps:
- Fill all the fixture traps with water
- Plug the building drain at the front main cleanout
- Burn oily rags or use other substance to produce odorous thick smoke with the help of a smoke machine
- Introduce the smoke into the system with an air pump
- When smoke appears at the opening on the roof, seal the opening
- Build up a pressure of 1 inch of water column and maintain within the system for 15 minutes
- Detect the leaks in the system by using the smell sense. Avoid any of the smoke odors on the clothing

Locating leaks with peppermint test

To locate leaks in sanitary drainage and vent systems using a peppermint test follow these steps:
- Pour two ounces of peppermint oil into each stack and the building drain making sure that fixture and floor drain traps are filled with water and the building drain is plugged at the front main cleanout
- Plug the stack and building drain openings
- Insert the manometer hose through the trap seal of a water closet bowl and blow water out of the hose
- Fill the manometer with water to the 0 mark on the ruler
- Attach the hose to the manometer and set the manometer to rest on the open closet seal
- Blow air into the system using a hose inserted through the trap of a water closet. Blow air till it reaches and maintains a 1-inch differential of pressure
- If the manometer does not hold the pressure, there are leaks in the system that can be found using the sense of smell

Testing pressure-reducing valves

Pressure reducing valves should be inspected regularly to ensure that the valves are maintained in operating condition. A partial flow test adequate to move the valve from its seat should be conducted annually. A full flow test should be conducted on each pressure-reducing valve at 5-year intervals. Results should be compared to the previous test results.

The static and residual inlet and outlet pressures of the valve should be recorded. Results should be compared with the manufacturer's performance curves to determine if the valve is operating properly. If there are any changes in test results, the condition of the valve should be evaluated and adjustments should be made as needed. If adjustments are necessary, they should be made in accordance with the manufacturer's recommendations.

Freeing jammed garbage disposal

The following are the steps to freeing a jammed garbage disposal:
- Disconnect the pipe leaving the disposal from the drain system piping under the sink or from the P-trap under the sink
- Attach the rubber test cap to the end of this pipe and tighten the clamp
- Place an empty five-gallon bucket under the end of this pipe
- Fill the disposal with warm water. Stop running the water as soon as the level reaches the top of the chrome strainer basket in the bottom of the sink that connects to the disposal

- Add one-half cup of powdered oxygen bleach to the garbage disposal filled with water and leave it for an hour
- Loosen the clamp on the rubber test cap and let the water into the bucket. Reconnect the disposer to the plumbing drain, insert the stopper into the sink, fill the sink with warm, soapy water, remove the stopper and turn on the disposer

Pressure reducing valve adjustment

Pressure reducing valves are fitted with several different types of pressure adjustment screws, such as a standard adjusting screw, a T-handle screw, a hand-wheel screw, and a tamper-proof cap. Pressure adjustment can be accomplished by turning the screw clockwise or counterclockwise. To increase the delivery pressure, turn the screw clockwise. To reduce the delivery pressure, turn the screw counterclockwise.

When the screw is turned clockwise, the adjusting spring acts against the diaphragm assembly and moves the internal valve seat to the open position. When high inlet pressure is applied, it flows into the regulator, through the open seat, up under the diaphragm, and on through the outlet. The outlet pressure builds up under the diaphragm to the adjusted setting. The downward adjusting spring pressure is overcome and the valve seat closes to maintain the required delivery pressure.

Ultrasonic test

A leaking steam system is a cause of energy waste. Steam may be lost through uninsulated valves, flanges, sections of steam pipe, or through high back pressure in condensate lines caused by blowing traps. To prevent the waste of energy, it is necessary to check piping and devices for leakage. The following is a procedure for testing steam pressure reducing valves with an ultrasonic device:
- Touch the ultrasonic instrument upstream of the valve or trap and reduce the sensitivity of the detector until the meter reads 50. If it is necessary to hear the specific sound quality of the fluid, tune the frequency until the sound becomes clear
- Touch downstream of the valve or trap and compare intensity levels. For traps, compare sound pattern levels

If the sound level is louder downstream, fluid is passing through. If the sound level is low, the valve or trap is closed.

Replacing a thermocouple

Follow these steps to replace a thermocouple:
- Turn off the gas supply to water heater. Remove outer and inner doors
- Disconnect burner assembly from the gas control
- Remove burner assembly from the combustion chamber
- Remove the old thermocouple from the bracket
- Install the new thermocouple. Position the thermocouple tip so that the pilot flame heats the top ½" of the tip
- Replace burner assembly in combustion chamber
- Tighten the main burner supply tube, pilot supply tube, and thermocouple connection to the gas control valve. Turn the thermocouple no more than ¼ turn beyond hand tight
- Turn on the gas supply. Check main and pilot supply tube at the gas control valve for leaks with a soapy water solution
- Light the pilot
- Check that ½" of the thermocouple tip is positioned in the flame
- Replace inner and outer doors

Testing a thermocouple

If the water heater does not remain lit, but the pilot light relights and goes out when the button is released, and the thermocouple is tight in the gas valve, the thermocouple test is required. Follow these steps to perform the test:
- Disconnect the thermocouple from the thermostat
- Using a multimeter with alligator clip leads attach the red lead to the body (copper part) of the thermocouple
- Attach the black lead to the end (silver part) of the thermocouple that connects to the thermostat
- Follow the manufacturer's instructions to light the pilot and watch the voltage readings on the multimeter. After 45 seconds the meter should read 12 millivolts or more

If the voltage is higher than 12 millivolts, replace the gas valve. If the voltage is lower than 12 millivolts, replace the thermocouple.

Testing backflow preventers

The following are the National Plumbing Code requirements for testing backflow preventers:
- Devices designed for field-testing should be tested before the final inspection of the initial installation and once a year after the initial installation. The field test procedures should comply with ANSI/ASSE 5010
- Testing and repair should be performed by certified individuals approved by an agency acceptable to the Administrative Authority. Certification for testing

should comply with ANSI/ASSE 5000. Certification for repair should comply with ANSI/ASSE 5030
- Copies of the initial installation test reports should be sent to the Administrative Authority and the water supplier. Copies of annual test reports should be sent to the water supplier
- In areas where a continuous water supply cannot be interrupted for the periodic testing of a backflow prevention device, multiple backflow prevention devices or other means of maintaining a continuous supply should be provided

Draft test

If the water heater does not remain lit, but the pilot light relights and stays lit until call for heat is satisfied, the draft test is required. Follow these steps to perform the test:
- Pass a match or smoke around the draft hood with the heater (main burner) in operation. It should draw under the draft hood into the vent pipe
- Turn on any air moving equipment, for example furnace, clothes dryer, attic fan, etc
- If it does not continue to draw under the draft hood with all the equipment on, shut off the water heater and increase combustion air or correct venting. Make sure to check the manufacturer's recommendations for requirements

Locating underground sewers

The following are the most common methods of locating underground sewers and metallic pipes:
- If one end of the sewer is accessible, the exact location of an underground sewer may be found by pounding on the ground surface or on a plank placed on the ground surface and listening at one end of the pipe for the variation of intensity of sound produced. The sewer should be located under the point where the maximum intensity of sound was produced
- If one end of the metallic pipe is accessible, its exact underground location may be found by using the pipe as the closed electric circuit. Electrical contacts are attached to two exposed points on the pipe and the electric circuit is completed. Earphones are used to listen to the electric current converted into sound

Maintenance of septic tanks

A septic tank is in normal condition if the top of the invert is visible above the liquid level. Septic tanks should be periodically cleaned to continue work properly. The system should not be allowed to operate until fixtures connected to it stopped draining. To remove accumulated sludge, septic tanks should be pumped out every 3-5 years depending on the type of the system and its efficiency.

Cleaning septic tanks

The following methods can be used to clean septic tanks:
- A long scoop-like device is used to remove solids from the bottom of the tank
- A pump or a reverse flushing mechanism is attached to the tank. The effluent is pumped into the truck and then the direction of the flow is reversed to help break up the sludge

Vent termination requirements

The following are the International and State Plumbing Codes requirements for vent termination:
- Vents should terminate into the atmosphere above the roof surface of the building
- The pipe should pass through a flashing and terminate in an approved vent cap installed according to the manufacturer's recommendations
- Gravity vent systems should terminate at least 5 feet above the highest vent collar
- Type B gas vents with approved vent caps may terminate at least 8 feet from a vertical wall or similar obstruction. All other type B vents should terminate at least 2 feet above the highest point where they pass through the roof and 2 feet higher than any section of the building within 10 feet
- Type L vents should terminate at least 2 feet above the roof and 4 feet higher than any section of the building extending at an angle of 45º upward

Locating collapsed underground pipes

If a pipe has collapsed, there may be increased local water-logging or an indication of a surface channel where the land has subsided in line with the pipe. However, wet areas do not necessarily indicate a broken pipe. If a wet patch is caused by a broken pipe, the break is usually just uphill of the wet patch. To locate collapsed underground pipes, follow these steps:
- Find an outlet of a specific drain. Drains might be curved near their outlet to slow the flow of water, so walk a few feet along the approximate line of the drain before attempting to locate it. Locate it again at a second point. Line up these two points to indicate the drain's alignment
- To follow the exact line of a pipe, insert a pointed metal pole or rod into the earth every few inches along it, until the drain is reached

Trailer hot water facility requirements

The following are the safety requirements for residential trailer hot water facilities:
- Residential trailers can be equipped only with approved automatic water heaters that have the minimum capacity of 5 gallons
- Hot water facilities should be used only at trailer parks where hot water piping is available for hook-up
- Residential trailer water heaters should be provided with an approved combination pressure and temperature relief valve
- Relief valves should be installed at 4 inches from the top of the water heater or less
- No more than 300-pound hydrostatic pressure test should be applied to water heater tanks
- Tank outlets should be provided with shutoff valves and drainpipes that lead outside of the trailer

Venting water heater requirements

The following are the International and State Plumbing Codes requirements for venting water heaters:
- Venting systems may consist of chimneys, type B and type L vents, and plastic pipe. All systems must be approved and installed according to the manufacturer's recommendations
- Vent design should allow a positive flow adequate to convey combustion products into the atmosphere
- Plastic venting materials may be used for condensing appliances that cool fuel gases to the nearly dew point and produce low-vent gas temperatures
- Unused opening in a vent system must be capped or closed to prevent entrance of foreign material
- Water heaters that may be converted to the use of solid and liquid fuel should not be vented with type B vents
- Manually operated dampers should not be installed in chimneys, vent, or chimney-vent connectors of fuel-fired water heaters

Hot water safety issues

The following are the safety issues related with domestic hot water:
- Hot water at temperatures higher than 131°F present a serious health hazard, therefore all hot water pipes must be properly insulated, and temperature-pressure relief valves should be installed according to the manufacturer's recommendations to prevent burns
- Hot water becomes more corrosive when the temperature rises every 20°F, therefore corrosion-resistance piping should be used for installation on water heaters

- 115 -

- Hot water at temperatures between 110° and 120°F serves as favorable environment for the legionnella bacterium, therefore it is important to maintain the hot water piping in clean condition

Installation of hot water recirculation system can prevent most of the safety-related problems.

Water heater safety requirements

The following are the safety requirements for water heaters:
- An anti-siphoning device should be installed to prevent siphoning of water from a water heater or a tank. For example, a cold-water dip tube with a hole at the top or a vacuum-relief valve in the cold water supply above the top of the water heater can be used
- Automatically controlled water heaters should be equipped with the energy cutoff valve to prevent the overheating of water (reaching the temperature of 210°F)
- Temperature-and-pressure relief valves are required for installation on any type of water heater
- Electric heaters should have electrical disconnect switch installed in close proximity to the water heater

Causes of back siphonage

The main cause of back siphonage from drains through water pipes is the failure or a significant reduction of water pressure. Reduction of water pressure, in its turn, may be caused by failures outside or inside the building. For example, some of the outside causes may be sudden flushing of street mains, fire engine pumping, breaks in mains, or improper closing of street valves. The inside causes may be meters, strainers, pressure regulators, undersized piping, and adding excessive number of fixtures to the existing pipes.

Prevention
Some of the prevention methods of back siphonage through water pipes include:
- installing a vacuum breaker in toilets
- marking pipes carrying unsafe water
- labeling valves and installing check valves
- checking overhead drain lines for leaks

Pipe support requirements

The following are the Plumbing Code requirements for pipe support:
- Pipe hangers should be compatible with the pipe they are supporting and should be made of the same material in order not to have a detrimental effect on the pipe, such as corrosion
- Plastic, or plastic-coated hangers, may be used with all types of pipes except for pipes that carry hot liquid that might melt the plastic
- Hangers must be securely attached to the pipe and to the member holding the hanger
- Hangers must support both horizontal and vertical piping
- The base of each cast-iron stack must be supported
- All flexible couplings installed on the pipe must be supported to prevent loosening of the connection

Bidet backflow prevention

The National Standard Plumbing Code requires that bidets with integral flushing rims should have a vacuum breaker assembly on the mixed water supply to the fixture. Bidets without flushing rims should have an over-the-rim supply fitting providing the air gap required by Chapter 10 of NSPC.

NSPC approved backflow preventer

The following is a description of an NSPC approved backflow preventer:
- A dual check backflow preventer with intermediate atmospheric vent (1/2" - 3/4") should be installed at referenced cross-connections
- The valve should feature stainless steel and rubber internals protected by an integral strainer
- The primary check should be rubber to rubber seated, backed by the secondary check with rubber to metal seating
- The device should be approved by ASSE under Standard 1012

<u>Prevention devices</u>
The following are the National Standard Plumbing Code requirements for backflow prevention devices:

- All backflow prevention devices should be accessible. Devices with atmospheric vents should not be installed in pits, vaults, or other potentially submerged locations. Vacuum breakers and other devices with vents to atmosphere should not be located within fume hoods
- Atmospheric vacuum breakers should be installed at least 6 inches above the flood level rim or highest point of discharge of the fixture being served. Deck-mounted and pipe-applied vacuum breakers and vacuum breakers within equipment, machinery and fixtures should be installed in accordance with manufacturer's instructions with the critical level no less than 1 inch above the flood level rim. These devices should be installed on the discharge side of the last control valve to the fixture and no shutoff valve or faucet should be installed downstream of the vacuum breaker. Vacuum breakers on urinals should be installed with the critical level 6 inches above the flood level rim

Flush tank backflow prevention

The following are the National Standard Plumbing Code requirements for flush tank backflow prevention:

- Flush tanks should have ball cocks or other means to refill the tank after each discharge and to shut off the water supply when the tank reaches the proper operating level. Ball cocks should be the anti-siphon type and comply with ANSI/ASSE 1002
- The seat of the tank flush valve should be at least 1 inch above the flood level rim of the fixture bowl
- The flush valve should close tightly if the tank is flushed when the fixture drain is clogged or partly restricted, so that water will not spill continuously over the rim of the bowl or backflow from the bowl to the flush tank
- Flush tanks should include a means of overflow into the fixture served with sufficient capacity to prevent the tank from overflowing when the ball cocks fully open

Sink faucet backflow prevention

The National Standard Plumbing Code requires that bidet sink faucets with a hose thread or other means of attaching a hose to the outlet should be protected from back-siphonage by either an integral vacuum breaker, an atmospheric vacuum breaker attached to the outlet, or pressure-type vacuum breakers on the fixture supply lines. The flow rate for kitchen faucets should not exceed the rate specified in ASNI/ASME A112.18.1M.

Approved anti-siphon vacuum breaker

The following is a description of an NSPC approved anti-siphon vacuum breaker:
- An atmospheric-type anti-siphon vacuum breaker (1/4" - 3") should be installed in places indicated on the plans to prevent the back-siphonage of contaminated water.
- This device should not be used under continuous pressure or where there is a possibility that a back pressure condition may develop
- The device shall meet the requirements of ASSE Standard 1001, ANSI A112.1.1 and CSA B64

NSPC required backflow prevention devices

The following are backflow prevention devices required by the National Standard Plumbing Code:
- Low Hazard - Back Siphonage - Intermittent Pressure
- Air gap, Atmospheric vacuum breaker
- Hose connection vacuum breaker
- Any backflow protection device approved for protection against continuous pressure back-siphonage
- Low Hazard - Back-Siphonage - Continuous Pressure
- Pressure-type vacuum breaker, Spill-proof vacuum breaker (SVB) Double check with intermediate atmospheric vent, Double check valve assembly
- Reduced pressure backflow preventer assemblies
- Low Hazard - Back-Pressure
- Double check with intermediate atmospheric vent, Double check valve assembly
- Reduced pressure backflow preventer assemblies
- High Hazard - Back-Siphonage
- Pressure-type vacuum breaker, Spill-proof vacuum breaker (SVB)
- Reduced pressure backflow preventer assemblies

<u>Installation required devices</u>
An atmospheric vacuum breaker is installed in an upright position with no valves downstream; the distance above the downstream piping and flood level rim of receptor should be no less than 6 inches. A spill-proof vacuum breaker (SVB) is installed in an upright position; the distance above the downstream piping and flood level rim of receptor should be no less than 6 inches.

A double check valve backflow preventer is installed in horizontal position with valves downstream; the distance above the downstream piping and flood level rim of receptor should be no less than 12 inches.

A pressure vacuum breaker is installed in an upright position with valves downstream; the distance above the downstream piping and flood level rim of receptor should be no less than 12 inches.

A reduced pressure backflow preventer is installed in horizontal position with valves downstream; the bottom clearance should be no less than 1 foot.

Potable water supply backflow prevention

The following are the National Standard Plumbing Code requirements for private potable water supply backflow prevention:
- Private potable water supplies such as wells, cisterns, lakes, streams, etc., should require the same backflow protection that is required for a public potable water supply
- Cross connection between a private potable water supply and a public potable water supply should not be made unless specifically approved by the appropriate administrative authority
- Interconnections between private supplies are prohibited because it is impossible to continuously monitor the water quality and potability of the private supply. Such supplies should be isolated from the public supply and properly tagged in conformance with section 10.2

Safe working habits

The following are the safe working habits:
- Wear safety equipment at all times
- Strictly observe safety rules at each location
- Anticipate and be aware of potential dangers in each situation
- Maintain tools in good working condition

Safe dressing habits

The following are the safe dressing habits:
- Do not wear easily flammable clothes
- Do not wear clothes with loose parts, such as wide sleeves, ties, etc
- Do not wear jewelry that can be caught in the machinery
- Wear protective gloves while handling hot or cold pipes and fittings
- Wear heavy protective boots that cannot be punctured with nails
- Do not wear shoes with loose shoelaces; always tighten shoelaces
- Wear protective hardhat when appropriate

Electric tool safety rules

The following are the safety rules of handling electric tools:
Always use a three-prong plug for any electric tool.

- Before using the tool, carefully read the instruction manual that came with it
- Before working with the tool, make sure it is properly grounded. In many cases, Occupational Safety and Health Administration (OSHA) requires ground fault circuit interrupters
- Always use sufficient length extension cord in order to avoid burning out a motor or cause other damage to equipment
- Make sure the extension cord does not run through water or any area where it can be accidentally cut, kinked, or run over
- While plugging in the extension cord, first hook it up to the equipment, and then plug it into the main electrical outlet
- Always coil up the extension cord after each use and store in a dry area

Hand tools safety rules

The following are the safety rules of handling hand tools:
- Always use the right tool for the job. Avoid awkward replacements
- Before using the tool, carefully read the instruction manual that came with it
- Maintain the tool in clean condition. Wipe after each use. Clean thoroughly at regular intervals
- Keep tools in good operating condition. Make sure chisels and saw blades are sharp, mushroomed heads are ground smooth, pipe wrenches are free of debris, the teeth are clean, etc.
- Do not carry small tools, especially sharp tools, in your pocket while working on ladders, scaffolding, etc. Tools might fall out and cause a serious injury to you or other workers

Trench and ditch safety rules

The following are the safety rules of working in trenches and ditches:
- While digging, pay careful attention to the underground utilities
- Make sure people are not standing on the top edge of the trench while there are workers in the trench
- Shore trenches deeper than 4 feet
- While digging, throw dirt at least 2 feet away from the walls of the trench
- When digging, pay careful attention to the water that might leak into the trench to avoid a cave-in. Special attention should be paid to the areas with a high water table
- Never work alone in a trench. Have someone nearby to call for help in case of an accident

- Keep a ladder in the trench in case you need to exit quickly
- Be aware and anticipate potential dangers. Remember that any heavy load truck might cause a cave-in

Grinder operation safety rules

The following are the safety rules of operating a grinder:
- Before using the grinder, carefully read the instruction manual that came with it
- Do not wear clothes with loose parts, such as wide sleeves, ties, etc
- Do not wear jewelry that can be caught in the machine
- Wear protective glasses or goggles at all times
- Do not wear gloves while operating the grinder to avoid getting caught in the machine
- Once you are finished operating the grinder, shut it off immediately. Avoid idle running of the machine
- Use the work rest to support and guide the tool
- Never remove guards from a grinder

Trench sloping and shoring safety

The following are the safety requirements for trench sloping and shoring:
- Trenches deeper than 5' in hard or compact soil must be shored or sloped to the angle of repose of the soil. The angle of repose of some soil types are: solid rock shale - 90º, compacted angular gravels - 63º, average soils - 45º, compacted sharp sand - 33º, rounded loose sand - 26º
- Trenches no deeper than 5' in loose sand, wet soil, or old fill must be shored or sloped to the angle of repose of the soil
- Portable trench boxes or sliding trench shields may be used instead of shoring or sloping if they provide equal or better protection compared with shoring
- All trench shoring must be done with high quality materials and designed by a qualified person
- All slopes, shoring, and grading should be inspected daily and after every rainstorm

Trench failure

A trench may fail in the following ways:
- the side walls may topple in
- the bottom of the trench walls may slide in
- the bottom of the trench may heave up due to the pressure of the weight of the side walls

- surface water may boil up into the bottom
- tension cracks may form and cause subsequent sliding and toppling
- subsidence and bulging may occur due to stress in unsupported soil mass

Prevention

The following can be used to prevent trench failure:

- Wood, steel, or concrete sheets forming a continuous line, placed in close contact and providing a wall to resist the lateral pressure of water, soil, or other materials
- Trench shield – steel plate and bracing welded or bolted together to support the walls of a trench from the ground level to the bottom. The trench shield can be moved as the work progresses

Ladder safety rules

The following are the safety rules of working on a ladder:

- Use a solid and level footing to set up the ladder
- Never use ladders that need repair
- Make sure all the steps of a stepladder are fully open and locked
- Place an extension ladder at least ¼ of its length away from the base of the building
- Always attach the extension ladder to the base of the building
- Extend the ladder at least 3 feet over the roof
- When climbing a ladder keep both hands free. Do not carry tools or materials
- Do not carry small tools, especially sharp tools, in your pocket
- Do not allow two workers to stand on the ladder designed for one person
- Keep the ladder clean, free from mud, grease, oil, etc

Working in rolling scaffolds safety

The following are the safety rules of working in rolling scaffolds:

- Never lay tools or materials on the floor of the scaffold. You might trip over them and fall. Tools and materials can also fall and injure workers underneath the scaffold
- Never move the scaffold while you are standing on it to avoid falling
- Lock the wheels immediately after positioning the scaffold
- Keep the scaffold level to maintain a steady working platform
- Inspect all equipment before using. Never use any equipment that is damaged or deteriorated in any way
- Anticipate and be prepared for potential dangerous situations
- Never overload scaffolds
- Never use ladders or makeshift devices on top of scaffolds to increase the height

Plumbing shop safety rules

The following are the safety rules to be observed in a plumbing shop:
- Always wear safety glasses and leather shoes
- Always pick up objects with your knees bent and your back straight
- Never move a length of pipe alone
- Always ask for help when lifting heavy objects
- Never run in the shop
- Do not throw objects
- Do not look at the arc on the electric welding machine
- Do not strike hardened materials together
- Always use tools only for their intended purpose
- Keep the fire equipment accessible
- Report injuries immediately
- Grind burrs off of chisels
- Always clean up after completing a job

Fire prevention

The following are the safety rules to prevent fire:
- Always keep fire extinguishers handy
- Make sure the extinguisher is full
- Familiarize yourself with the operation of an extinguisher
- Disconnect and bleed all hoses and regulators used in welding, brazing, soldering, etc.
- Make sure to store containers of acetylene, propane, oxygen, and other flammable substances in an upright position
- Never use propane torches near flammable equipment
- Always operate all air, acetylene, welding, soldering, and other related equipment in accordance to the manufacturer's instructions
- Make sure the fire exits are not blocked
- Maintain fire alarms in proper operating condition
- Be aware and anticipate potential dangers

Volumes

Small volumes are expressed in cubic inches. According to the US Bureau of Standards, there are 231 cubic inches in 1 gallon. Larger volumes are expressed in cubic feet. There are 7.5 gallons in one cubic foot. To calculate a volume of a water tank in inches and weight of water, do the following:
- Divide the volume in cubic inches by 231. The result of the computation will be the volume of the tank in gallons
- Multiply the volume in gallons by 8.33. The result of the computation will be the weight of water

Earth and concrete volumes are expressed in cubic yards. First the volume should be calculated in cubic feet, and then converted into cubic yards by dividing the volume by 27.

Offset and diagonal

Offset is a distance between the centerline of two parallel pipes that may be positioned horizontally or vertically. If the pipes are connected by fit other than 90 degrees, the offset is called diagonal.

The offset and the diagonal are proportionally dependent on each other. If one is changed, the other will be changed too. The ration representing this change is called a constant. The offset can be measured on the job and therefore is always known. The constants remain equal for any angle of the diagonal. The diagonal can be found by multiplying the offset by a constant.

Rise and run

Rise and run are terms used to indicate the distance between two parallel vertical and horizontal pipes respectively..

Calculating volume of a cylinder

To calculate the volume of a cylindrical tank, the following formula should be used:
- V=0.7854d2h where d is diameter measured in inches and h is height measured in feet

Calculating volume of a sphere

To calculate the volume of a spherical tank, the following formulas should be used:
- V=0.5236d3 – volume of a sphere or V=0.2618d3 – volume of a half-sphere where d is diameter.\

Calculating volume of a segment

Segment volume calculation is used when cylindrical tanks are installed on the side. The volume equals the area multiplied by the length of the tank.

To calculate the area of a segment, the following formula should be used:

$$V = \frac{4h^2}{2}\sqrt{\frac{d}{h}} - 0.608$$

Volume of a standard size pipe

The volume of water held in a particular length of a standard weight pipe can be found by multiplying the length of the pipe in feet and the number of gallons held in one foot of a particular size pipe. The number of gallons that is held in one foot of a particular size pipe is found in tables containing weights and capacities for standard pipe sizes.

For example, if it is known that the pipe length is 18 feet and the pipe size is 3 inches, we can determine that the number of gallons held in one foot of this size pipe is 0.384 gallons per linear foot. To calculate the volume of water held in 18 feet of pipe, it is necessary to multiply 0.384 gallons per linear foot by 18 feet. The result is 6.912 gallons.

Explain how to calculate circumference, diameter, and area of a circle

Calculating circumference of circle

To calculate the circumference of a circle if the diameter is known, multiply the diameter by π (3.1416)

Calculating diameter of circle

To calculate the diameter of a circle if the circumference is known, divide the circumference by π (3.1416)

To calculate the diameter of a circle if the area is known, extract the square root of the area divided by π/4 (0.7854).

Calculating area of circle

To calculate the area of a circle if the diameter is known, multiply the square of a diameter by π/4 (0.7854)

To calculate the area of a sector of a circle if the radius is known, multiply the arc of the sector by half the radius.

To calculate the area of a ring if the diameter is known, calculate the area of the outside and the inside circles; then subtract the area of the inside circle from the area of the outside circle.

Calculating volumes of partially filled containers

Pools, linked tanks, heat exchangers, natural water reservoirs are examples of partially filled containers. Different materials fill the containers in different manner that complicates the calculations even more. For example, liquids maintain parallel

level with the ground while gases and solids do not. One strategy to determine a liquid volume of a partially filled container is to test a quantity of water supplied by a given fill line per minute and then filling the empty tank and making marks beside the sight glass as the water rises. To calculate a liquid volume of the empty part of a container, do the following:
- Calculate the total volume of the empty container
- Mark the level of water in the partially filled container
- Calculate the liquid volume of the filled part
- Subtract the liquid volume of the filled part from the total volume

Calculating fitting allowance

The following are the steps for calculating fitting allowance:
- Select a short length of pipe
- Measure the length and write it down
- Place a fitting on the measured piece of pipe. If the pipe is threaded, tighten the fitting with the same wrench and force that will be used for the actual fitting
- Place the pipe and fitting assembly in a vertical position on a flat surface and measure the distance from the surface to the centerline of the fitting
- Subtract the measurement of the first length of the pipe from the last measurement. The resulting number is the fitting allowance

Calculating weight of water

One cubic inch of water weighs 0.0361 pound. For example, it is necessary to calculate the weight of a column of water 12" high and 1" in diameter. Multiply the height of water by the weight of 1 cubic inch of water. 12" X 0.0361lb = 0.4332 lb. The formula is $W = h \times d \times 0.0361$.

Calculating weight and pressure of water

A 1-foot water column has pressure of 0.4332 psi. For example, it is necessary to calculate water pressure at the city water main required to fill a house tank located on top of a 6-story building. The inlet to the water tank is located at an elevation of 110 feet above the city water main. Since it is known that 1-foot water column has pressure of 0.4332 psi, the water pressure required is 110' X 0.4332 psi = 47.652 psi.

Secret Key #1 - Time is Your Greatest Enemy

Pace Yourself

Wear a watch. At the beginning of the test, check the time (or start a chronometer on your watch to count the minutes), and check the time after every few questions to make sure you are "on schedule."

If you are forced to speed up, do it efficiently. Usually one or more answer choices can be eliminated without too much difficulty. Above all, don't panic. Don't speed up and just begin guessing at random choices. By pacing yourself, and continually monitoring your progress against your watch, you will always know exactly how far ahead or behind you are with your available time. If you find that you are one minute behind on the test, don't skip one question without spending any time on it, just to catch back up. Take 15 fewer seconds on the next four questions, and after four questions you'll have caught back up. Once you catch back up, you can continue working each problem at your normal pace.

Furthermore, don't dwell on the problems that you were rushed on. If a problem was taking up too much time and you made a hurried guess, it must be difficult. The difficult questions are the ones you are most likely to miss anyway, so it isn't a big loss. It is better to end with more time than you need than to run out of time.

Lastly, sometimes it is beneficial to slow down if you are constantly getting ahead of time. You are always more likely to catch a careless mistake by working more slowly than quickly, and among very high-scoring test takers (those who are likely to have lots of time left over), careless errors affect the score more than mastery of material.

Secret Key #2 - Guessing is not Guesswork

You probably know that guessing is a good idea. Unlike other standardized tests, there is no penalty for getting a wrong answer. Even if you have no idea about a question, you still have a 20-25% chance of getting it right.

Most test takers do not understand the impact that proper guessing can have on their score. Unless you score extremely high, guessing will significantly contribute to your final score.

Monkeys Take the Test

What most test takers don't realize is that to insure that 20-25% chance, you have to guess randomly. If you put 20 monkeys in a room to take this test, assuming they answered once per question and behaved themselves, on average they would get 20-25% of the questions correct. Put 20 test takers in the room, and the average will be much lower among guessed questions. Why?
1. The test writers intentionally write deceptive answer choices that "look" right. A test taker has no idea about a question, so he picks the "best looking" answer, which is often wrong. The monkey has no idea what looks good and what doesn't, so it will consistently be right about 20-25% of the time.
2. Test takers will eliminate answer choices from the guessing pool based on a hunch or intuition. Simple but correct answers often get excluded, leaving a 0% chance of being correct. The monkey has no clue, and often gets lucky with the best choice.

This is why the process of elimination endorsed by most test courses is flawed and detrimental to your performance. Test takers don't guess; they make an ignorant stab in the dark that is usually worse than random.

$5 Challenge

Let me introduce one of the most valuable ideas of this course—the $5 challenge:

You only mark your "best guess" if you are willing to bet $5 on it.
You only eliminate choices from guessing if you are willing to bet $5 on it.

Why $5? Five dollars is an amount of money that is small yet not insignificant, and can really add up fast (20 questions could cost you $100). Likewise, each answer choice on one question of the test will have a small impact on your overall score, but it can really add up to a lot of points in the end.

The process of elimination IS valuable. The following shows your chance of guessing it right:

If you eliminate wrong answer choices until only this many remain:	Chance of getting it correct:
1	100%
2	50%
3	33%

However, if you accidentally eliminate the right answer or go on a hunch for an incorrect answer, your chances drop dramatically—to 0%. By guessing among all the answer choices, you are GUARANTEED to have a shot at the right answer.

That's why the $5 test is so valuable. If you give up the advantage and safety of a pure guess, it had better be worth the risk.

What we still haven't covered is how to be sure that whatever guess you make is truly random. Here's the easiest way:

Always pick the first answer choice among those remaining.

Such a technique means that you have decided, **before you see a single test question**, exactly how you are going to guess, and since the order of choices tells you nothing about which one is correct, this guessing technique is perfectly random.

This section is not meant to scare you away from making educated guesses or eliminating choices; you just need to define when a choice is worth eliminating. The $5 test, along with a pre-defined random guessing strategy, is the best way to make sure you reap all of the benefits of guessing.

Secret Key #3 - Practice Smarter, Not Harder

Many test takers delay the test preparation process because they dread the awful amounts of practice time they think necessary to succeed on the test. We have refined an effective method that will take you only a fraction of the time.

There are a number of "obstacles" in the path to success. Among these are answering questions, finishing in time, and mastering test-taking strategies. All must be executed on the day of the test at peak performance, or your score will suffer. The test is a mental marathon that has a large impact on your future.

Just like a marathon runner, it is important to work your way up to the full challenge. So first you just worry about questions, and then time, and finally strategy:

Success Strategy

1. Find a good source for practice tests.
2. If you are willing to make a larger time investment, consider using more than one study guide. Often the different approaches of multiple authors will help you "get" difficult concepts.
3. Take a practice test with no time constraints, with all study helps, "open book." Take your time with questions and focus on applying strategies.
4. Take a practice test with time constraints, with all guides, "open book."
5. Take a final practice test without open material and with time limits.

If you have time to take more practice tests, just repeat step 5. By gradually exposing yourself to the full rigors of the test environment, you will condition your mind to the stress of test day and maximize your success.

Secret Key #4 - Prepare, Don't Procrastinate

Let me state an obvious fact: if you take the test three times, you will probably get three different scores. This is due to the way you feel on test day, the level of preparedness you have, and the version of the test you see. Despite the test writers' claims to the contrary, some versions of the test WILL be easier for you than others.

Since your future depends so much on your score, you should maximize your chances of success. In order to maximize the likelihood of success, you've got to prepare in advance. This means taking practice tests and spending time learning the information and test taking strategies you will need to succeed.

Never go take the actual test as a "practice" test, expecting that you can just take it again if you need to. Take all the practice tests you can on your own, but when you go to take the official test, be prepared, be focused, and do your best the first time!

Secret Key #5 - Test Yourself

Everyone knows that time is money. There is no need to spend too much of your time or too little of your time preparing for the test. You should only spend as much of your precious time preparing as is necessary for you to get the score you need.

Once you have taken a practice test under real conditions of time constraints, then you will know if you are ready for the test or not.

If you have scored extremely high the first time that you take the practice test, then there is not much point in spending countless hours studying. You are already there.

Benchmark your abilities by retaking practice tests and seeing how much you have improved. Once you consistently score high enough to guarantee success, then you are ready.

If you have scored well below where you need, then knuckle down and begin studying in earnest. Check your improvement regularly through the use of practice tests under real conditions. Above all, don't worry, panic, or give up. The key is perseverance!

Then, when you go to take the test, remain confident and remember how well you did on the practice tests. If you can score high enough on a practice test, then you can do the same on the real thing.

General Strategies

The most important thing you can do is to ignore your fears and jump into the test immediately. Do not be overwhelmed by any strange-sounding terms. You have to jump into the test like jumping into a pool—all at once is the easiest way.

Make Predictions

As you read and understand the question, try to guess what the answer will be. Remember that several of the answer choices are wrong, and once you begin reading them, your mind will immediately become cluttered with answer choices designed to throw you off. Your mind is typically the most focused immediately after you have read the question and digested its contents. If you can, try to predict what the correct answer will be. You may be surprised at what you can predict.

Quickly scan the choices and see if your prediction is in the listed answer choices. If it is, then you can be quite confident that you have the right answer. It still won't hurt to check the other answer choices, but most of the time, you've got it!

Answer the Question

It may seem obvious to only pick answer choices that answer the question, but the test writers can create some excellent answer choices that are wrong. Don't pick an answer just because it sounds right, or you believe it to be true. It MUST answer the question. Once you've made your selection, always go back and check it against the question and make sure that you didn't misread the question and that the answer choice does answer the question posed.

Benchmark

After you read the first answer choice, decide if you think it sounds correct or not. If it doesn't, move on to the next answer choice. If it does, mentally mark that answer choice. This doesn't mean that you've definitely selected it as your answer choice, it just means that it's the best you've seen thus far. Go ahead and read the next choice. If the next choice is worse than the one you've already selected, keep going to the next answer choice. If the next choice is better than the choice you've already selected, mentally mark the new answer choice as your best guess.

The first answer choice that you select becomes your standard. Every other answer choice must be benchmarked against that standard. That choice is correct until proven otherwise by another answer choice beating it out. Once you've decided that no other answer choice seems as good, do one final check to ensure that your answer choice answers the question posed.

Valid Information

Don't discount any of the information provided in the question. Every piece of information may be necessary to determine the correct answer. None of the information in the question is there to throw you off (while the answer choices will certainly have information to throw you off). If two seemingly unrelated topics are discussed, don't ignore either. You can be confident there is a relationship, or it wouldn't be included in the question, and you are probably going to have to determine what is that relationship to find the answer.

Avoid "Fact Traps"

Don't get distracted by a choice that is factually true. Your search is for the answer that answers the question. Stay focused and don't fall for an answer that is true but irrelevant. Always go back to the question and make sure you're choosing an answer that actually answers the question and is not just a true statement. An answer can be factually correct, but it MUST answer the question asked. Additionally, two answers can both be seemingly correct, so be sure to read all of the answer choices, and make sure that you get the one that BEST answers the question.

Milk the Question

Some of the questions may throw you completely off. They might deal with a subject you have not been exposed to, or one that you haven't reviewed in years. While your lack of knowledge about the subject will be a hindrance, the question itself can give you many clues that will help you find the correct answer. Read the question carefully and look for clues. Watch particularly for adjectives and nouns describing difficult terms or words that you don't recognize. Regardless of whether you completely understand a word or not, replacing it with a synonym, either provided or one you more familiar with, may help you to understand what the questions are asking. Rather than wracking your mind about specific detailed information concerning a difficult term or word, try to use mental substitutes that are easier to understand.

The Trap of Familiarity

Don't just choose a word because you recognize it. On difficult questions, you may not recognize a number of words in the answer choices. The test writers don't put "make-believe" words on the test, so don't think that just because you only recognize all the words in one answer choice that that answer choice must be correct. If you only recognize words in one answer choice, then focus on that one. Is it correct? Try your best to determine if it is correct. If it is, that's great. If not, eliminate it. Each word and answer choice you eliminate increases your chances of getting the question correct, even if you then have to guess among the unfamiliar choices.

Eliminate Answers

Eliminate choices as soon as you realize they are wrong. But be careful! Make sure you consider all of the possible answer choices. Just because one appears right, doesn't mean that the next one won't be even better! The test writers will usually put more than one good answer choice for every question, so read all of them. Don't worry if you are stuck between two that seem right. By getting down to just two remaining possible choices, your odds are now 50/50. Rather than wasting too much time, play the odds. You are guessing, but guessing wisely because you've been able to knock out some of the answer choices that you know are wrong. If you are eliminating choices and realize that the last answer choice you are left with is also obviously wrong, don't panic. Start over and consider each choice again. There may easily be something that you missed the first time and will realize on the second pass.

Tough Questions

If you are stumped on a problem or it appears too hard or too difficult, don't waste time. Move on! Remember though, if you can quickly check for obviously incorrect answer choices, your chances of guessing correctly are greatly improved. Before you completely give up, at least try to knock out a couple of possible answers. Eliminate what you can and then guess at the remaining answer choices before moving on.

Brainstorm

If you get stuck on a difficult question, spend a few seconds quickly brainstorming. Run through the complete list of possible answer choices. Look at each choice and ask yourself, "Could this answer the question satisfactorily?" Go through each answer choice and consider it independently of the others. By systematically going through all possibilities, you may find something that you would otherwise overlook. Remember though that when you get stuck, it's important to try to keep moving.

Read Carefully

Understand the problem. Read the question and answer choices carefully. Don't miss the question because you misread the terms. You have plenty of time to read each question thoroughly and make sure you understand what is being asked. Yet a happy medium must be attained, so don't waste too much time. You must read carefully, but efficiently.

Face Value

When in doubt, use common sense. Always accept the situation in the problem at face value. Don't read too much into it. These problems will not require you to make huge leaps of logic. The test writers aren't trying to throw you off with a cheap trick. If you have to go beyond creativity and make a leap of logic in order to have an answer choice answer the question, then you should look at the other answer choices. Don't overcomplicate the problem by creating theoretical relationships or explanations that will warp time or space. These are normal problems rooted in reality. It's just that the applicable relationship or explanation may not be readily apparent and you have to figure things out. Use your common sense to interpret anything that isn't clear.

Prefixes

If you're having trouble with a word in the question or answer choices, try dissecting it. Take advantage of every clue that the word might include. Prefixes and suffixes can be a huge help. Usually they allow you to determine a basic meaning. Pre- means before, post- means after, pro - is positive, de- is negative. From these prefixes and suffixes, you can get an idea of the general meaning of the word and try to put it into context. Beware though of any traps. Just because con- is the opposite of pro-, doesn't necessarily mean congress is the opposite of progress!

Hedge Phrases

Watch out for critical hedge phrases, led off with words such as "likely," "may," "can," "sometimes," "often," "almost," "mostly," "usually," "generally," "rarely," and "sometimes." Question writers insert these hedge phrases to cover every possibility. Often an answer choice will be wrong simply because it leaves no room for exception. Unless the situation calls for them, avoid answer choices that have definitive words like "exactly," and "always."

Switchback Words

Stay alert for "switchbacks." These are the words and phrases frequently used to alert you to shifts in thought. The most common switchback word is "but." Others include "although," "however," "nevertheless," "on the other hand," "even though," "while," "in spite of," "despite," and "regardless of."

New Information

Correct answer choices will rarely have completely new information included. Answer choices typically are straightforward reflections of the material asked about and will directly relate to the question. If a new piece of information is included in an answer choice that doesn't even seem to relate to the topic being asked about, then that answer choice is likely incorrect. All of the information needed to answer the question is usually provided for you in the question. You should not have to make guesses that are unsupported or choose answer choices that require unknown information that cannot be reasoned from what is given.

Time Management

On technical questions, don't get lost on the technical terms. Don't spend too much time on any one question. If you don't know what a term means, then odds are you aren't going to get much further since you don't have a dictionary. You should be able to immediately recognize whether or not you know a term. If you don't, work with the other clues that you have—the other answer choices and terms provided— but don't waste too much time trying to figure out a difficult term that you don't know.

Contextual Clues

Look for contextual clues. An answer can be right but not the correct answer. The contextual clues will help you find the answer that is most right and is correct. Understand the context in which a phrase or statement is made. This will help you make important distinctions.

Don't Panic

Panicking will not answer any questions for you; therefore, it isn't helpful. When you first see the question, if your mind goes blank, take a deep breath. Force yourself to mechanically go through the steps of solving the problem using the strategies you've learned.

Pace Yourself

Don't get clock fever. It's easy to be overwhelmed when you're looking at a page full of questions, your mind is full of random thoughts and feeling confused, and the clock is ticking down faster than you would like. Calm down and maintain the pace that you have set for yourself. As long as you are on track by monitoring your pace, you are guaranteed to have enough time for yourself. When you get to the last few minutes of the test, it may seem like you won't have enough time left, but if you only have as many questions as you should have left at that point, then you're right on track!

Answer Selection

The best way to pick an answer choice is to eliminate all of those that are wrong, until only one is left and confirm that is the correct answer. Sometimes though, an answer choice may immediately look right. Be careful! Take a second to make sure that the other choices are not equally obvious. Don't make a hasty mistake. There are only two times that you should stop before checking other answers. First is when you are positive that the answer choice you have selected is correct. Second is when time is almost out and you have to make a quick guess!

Check Your Work

Since you will probably not know every term listed and the answer to every question, it is important that you get credit for the ones that you do know. Don't miss any questions through careless mistakes. If at all possible, try to take a second to look back over your answer selection and make sure you've selected the correct answer choice and haven't made a costly careless mistake (such as marking an answer choice that you didn't mean to mark). The time it takes for this quick double check should more than pay for itself in caught mistakes.

Beware of Directly Quoted Answers

Sometimes an answer choice will repeat word for word a portion of the question or reference section. However, beware of such exact duplication. It may be a trap! More than likely, the correct choice will paraphrase or summarize a point, rather than being exactly the same wording.

Slang

Scientific sounding answers are better than slang ones. An answer choice that begins "To compare the outcomes..." is much more likely to be correct than one that begins "Because some people insisted..."

Extreme Statements

Avoid wild answers that throw out highly controversial ideas that are proclaimed as established fact. An answer choice that states the "process should used in certain situations, if..." is much more likely to be correct than one that states the "process should be discontinued completely." The first is a calm rational statement and doesn't even make a definitive, uncompromising stance, using a hedge word "if" to provide wiggle room, whereas the second choice is a radical idea and far more extreme.

Answer Choice Families

When you have two or more answer choices that are direct opposites or parallels, one of them is usually the correct answer. For instance, if one answer choice states "x increases" and another answer choice states "x decreases" or "y increases," then those two or three answer choices are very similar in construction and fall into the same family of answer choices. A family of answer choices consists of two or three answer choices, very similar in construction, but often with directly opposite meanings. Usually the correct answer choice will be in that family of answer choices. The "odd man out" or answer choice that doesn't seem to fit the parallel construction of the other answer choices is more likely to be incorrect.

Special Report: How to Overcome Test Anxiety

The very nature of tests caters to some level of anxiety, nervousness, or tension, just as we feel for any important event that occurs in our lives. A little bit of anxiety or nervousness can be a good thing. It helps us with motivation, and makes achievement just that much sweeter. However, too much anxiety can be a problem, especially if it hinders our ability to function and perform.

"Test anxiety," is the term that refers to the emotional reactions that some test-takers experience when faced with a test or exam. Having a fear of testing and exams is based upon a rational fear, since the test-taker's performance can shape the course of an academic career. Nevertheless, experiencing excessive fear of examinations will only interfere with the test-taker's ability to perform and chance to be successful.

There are a large variety of causes that can contribute to the development and sensation of test anxiety. These include, but are not limited to, lack of preparation and worrying about issues surrounding the test.

Lack of Preparation

Lack of preparation can be identified by the following behaviors or situations:

Not scheduling enough time to study, and therefore cramming the night before the test or exam
Managing time poorly, to create the sensation that there is not enough time to do everything
Failing to organize the text information in advance, so that the study material consists of the entire text and not simply the pertinent information
Poor overall studying habits

Worrying, on the other hand, can be related to both the test taker, or many other factors around him/her that will be affected by the results of the test. These include worrying about:

Previous performances on similar exams, or exams in general
How friends and other students are achieving
The negative consequences that will result from a poor grade or failure

There are three primary elements to test anxiety. Physical components, which involve the same typical bodily reactions as those to acute anxiety (to be discussed below). Emotional factors have to do with fear or panic. Mental or cognitive issues concerning attention spans and memory abilities.

Physical Signals

There are many different symptoms of test anxiety, and these are not limited to mental and emotional strain. Frequently there are a range of physical signals that will let a test taker know that he/she is suffering from test anxiety. These bodily changes can include the following:

Perspiring
Sweaty palms
Wet, trembling hands
Nausea
Dry mouth
A knot in the stomach
Headache
Faintness
Muscle tension
Aching shoulders, back and neck
Rapid heart beat
Feeling too hot/cold

To recognize the sensation of test anxiety, a test-taker should monitor him/herself for the following sensations:

The physical distress symptoms as listed above
Emotional sensitivity, expressing emotional feelings such as the need to cry or laugh too much, or a sensation of anger or helplessness
A decreased ability to think, causing the test-taker to blank out or have racing thoughts that are hard to organize or control.

Though most students will feel some level of anxiety when faced with a test or exam, the majority can cope with that anxiety and maintain it at a manageable level. However, those who cannot are faced with a very real and very serious condition, which can and should be controlled for the immeasurable benefit of this sufferer.

Naturally, these sensations lead to negative results for the testing experience. The most common effects of test anxiety have to do with nervousness and mental blocking.

Nervousness

Nervousness can appear in several different levels:

The test-taker's difficulty, or even inability to read and understand the questions on the test

The difficulty or inability to organize thoughts to a coherent form
The difficulty or inability to recall key words and concepts relating to the testing questions (especially essays)
The receipt of poor grades on a test, though the test material was well known by the test taker

Conversely, a person may also experience mental blocking, which involves:

Blanking out on test questions
Only remembering the correct answers to the questions when the test has already finished.

Fortunately for test anxiety sufferers, beating these feelings, to a large degree, has to do with proper preparation. When a test taker has a feeling of preparedness, then anxiety will be dramatically lessened.

The first step to resolving anxiety issues is to distinguish which of the two types of anxiety are being suffered. If the anxiety is a direct result of a lack of preparation, this should be considered a normal reaction, and the anxiety level (as opposed to the test results) shouldn't be anything to worry about. However, if, when adequately prepared, the test-taker still panics, blanks out, or seems to overreact, this is not a fully rational reaction. While this can be considered normal too, there are many ways to combat and overcome these effects.

Remember that anxiety cannot be entirely eliminated, however, there are ways to minimize it, to make the anxiety easier to manage. Preparation is one of the best ways to minimize test anxiety. Therefore the following techniques are wise in order to best fight off any anxiety that may want to build.

To begin with, try to avoid cramming before a test, whenever it is possible. By trying to memorize an entire term's worth of information in one day, you'll be shocking your system, and not giving yourself a very good chance to absorb the information. This is an easy path to anxiety, so for those who suffer from test anxiety, cramming should not even be considered an option.

Instead of cramming, work throughout the semester to combine all of the material which is presented throughout the semester, and work on it gradually as the course goes by, making sure to master the main concepts first, leaving minor details for a week or so before the test.

To study for the upcoming exam, be sure to pose questions that may be on the examination, to gauge the ability to answer them by integrating the ideas from your texts, notes and lectures, as well as any supplementary readings.

If it is truly impossible to cover all of the information that was covered in that particular term, concentrate on the most important portions, that can be covered very well. Learn these concepts as best as possible, so that when the test comes, a goal can be made to use these concepts as presentations of your knowledge.

In addition to study habits, changes in attitude are critical to beating a struggle with test anxiety. In fact, an improvement of the perspective over the entire test-taking experience can actually help a test taker to enjoy studying and therefore improve the overall experience. Be certain not to overemphasize the significance of the grade - know that the result of the test is neither a reflection of self worth, nor is it a measure of intelligence; one grade will not predict a person's future success.

To improve an overall testing outlook, the following steps should be tried:

Keeping in mind that the most reasonable expectation for taking a test is to expect to try to demonstrate as much of what you know as you possibly can. Reminding ourselves that a test is only one test; this is not the only one, and there will be others.
The thought of thinking of oneself in an irrational, all-or-nothing term should be avoided at all costs.
A reward should be designated for after the test, so there's something to look forward to. Whether it be going to a movie, going out to eat, or simply visiting friends, schedule it in advance, and do it no matter what result is expected on the exam.

Test-takers should also keep in mind that the basics are some of the most important things, even beyond anti-anxiety techniques and studying. Never neglect the basic social, emotional and biological needs, in order to try to absorb information. In order to best achieve, these three factors must be held as just as important as the studying itself.

Study Steps

Remember the following important steps for studying:

Maintain healthy nutrition and exercise habits. Continue both your recreational activities and social pass times. These both contribute to your physical and emotional well being.
Be certain to get a good amount of sleep, especially the night before the test, because when you're overtired you are not able to perform to the best of your best ability.
Keep the studying pace to a moderate level by taking breaks when they are needed, and varying the work whenever possible, to keep the mind fresh instead of getting bored.
When enough studying has been done that all the material that can be learned has been learned, and the test taker is prepared for the test, stop studying and do

something relaxing such as listening to music, watching a movie, or taking a warm bubble bath.

There are also many other techniques to minimize the uneasiness or apprehension that is experienced along with test anxiety before, during, or even after the examination. In fact, there are a great deal of things that can be done to stop anxiety from interfering with lifestyle and performance. Again, remember that anxiety will not be eliminated entirely, and it shouldn't be. Otherwise that "up" feeling for exams would not exist, and most of us depend on that sensation to perform better than usual. However, this anxiety has to be at a level that is manageable.

Of course, as we have just discussed, being prepared for the exam is half the battle right away. Attending all classes, finding out what knowledge will be expected on the exam, and knowing the exam schedules are easy steps to lowering anxiety. Keeping up with work will remove the need to cram, and efficient study habits will eliminate wasted time. Studying should be done in an ideal location for concentration, so that it is simple to become interested in the material and give it complete attention. A method such as SQ3R (Survey, Question, Read, Recite, Review) is a wonderful key to follow to make sure that the study habits are as effective as possible, especially in the case of learning from a textbook. Flashcards are great techniques for memorization. Learning to take good notes will mean that notes will be full of useful information, so that less sifting will need to be done to seek out what is pertinent for studying. Reviewing notes after class and then again on occasion will keep the information fresh in the mind. From notes that have been taken summary sheets and outlines can be made for simpler reviewing.

A study group can also be a very motivational and helpful place to study, as there will be a sharing of ideas, all of the minds can work together, to make sure that everyone understands, and the studying will be made more interesting because it will be a social occasion.

Basically, though, as long as the test-taker remains organized and self confident, with efficient study habits, less time will need to be spent studying, and higher grades will be achieved.

To become self confident, there are many useful steps. The first of these is "self talk." It has been shown through extensive research, that self-talk for students who suffer from test anxiety, should be well monitored, in order to make sure that it contributes to self confidence as opposed to sinking the student. Frequently the self talk of test-anxious students is negative or self-defeating, thinking that everyone else is smarter and faster, that they always mess up, and that if they don't do well, they'll fail the entire course. It is important to decreasing anxiety that awareness is made of self talk. Try writing any negative self thoughts and then disputing them with a positive statement instead. Begin

Copyright © Mometrix Media. You have been licensed one copy of this document for personal use only. Any other reproduction or redistribution is strictly prohibited. All rights reserved.

self-encouragement as though it was a friend speaking. Repeat positive statements to help reprogram the mind to believing in successes instead of failures.

Helpful Techniques

Other extremely helpful techniques include:

Self-visualization of doing well and reaching goals
While aiming for an "A" level of understanding, don't try to "overprotect" by setting your expectations lower. This will only convince the mind to stop studying in order to meet the lower expectations.
Don't make comparisons with the results or habits of other students. These are individual factors, and different things work for different people, causing different results.
Strive to become an expert in learning what works well, and what can be done in order to improve. Consider collecting this data in a journal.
Create rewards for after studying instead of doing things before studying that will only turn into avoidance behaviors.
Make a practice of relaxing - by using methods such as progressive relaxation, self-hypnosis, guided imagery, etc - in order to make relaxation an automatic sensation.
Work on creating a state of relaxed concentration so that concentrating will take on the focus of the mind, so that none will be wasted on worrying.
Take good care of the physical self by eating well and getting enough sleep.
Plan in time for exercise and stick to this plan.

Beyond these techniques, there are other methods to be used before, during and after the test that will help the test-taker perform well in addition to overcoming anxiety.

Before the exam comes the academic preparation. This involves establishing a study schedule and beginning at least one week before the actual date of the test. By doing this, the anxiety of not having enough time to study for the test will be automatically eliminated. Moreover, this will make the studying a much more effective experience, ensuring that the learning will be an easier process. This relieves much undue pressure on the test-taker.

Summary sheets, note cards, and flash cards with the main concepts and examples of these main concepts should be prepared in advance of the actual studying time. A topic should never be eliminated from this process. By omitting a topic because it isn't expected to be on the test is only setting up the test-taker for anxiety should it actually appear on the exam. Utilize the course syllabus for laying out the topics that should be studied. Carefully go over the notes that were made in class, paying special attention to any of the issues that

- 145 -

the professor took special care to emphasize while lecturing in class. In the textbooks, use the chapter review, or if possible, the chapter tests, to begin your review.

It may even be possible to ask the instructor what information will be covered on the exam, or what the format of the exam will be (for example, multiple choice, essay, free form, true-false). Additionally, see if it is possible to find out how many questions will be on the test. If a review sheet or sample test has been offered by the professor, make good use of it, above anything else, for the preparation for the test. Another great resource for getting to know the examination is reviewing tests from previous semesters. Use these tests to review, and aim to achieve a 100% score on each of the possible topics. With a few exceptions, the goal that you set for yourself is the highest one that you will reach.

Take all of the questions that were assigned as homework, and rework them to any other possible course material. The more problems reworked, the more skill and confidence will form as a result. When forming the solution to a problem, write out each of the steps. Don't simply do head work. By doing as many steps on paper as possible, much clarification and therefore confidence will be formed. Do this with as many homework problems as possible, before checking the answers. By checking the answer after each problem, a reinforcement will exist, that will not be on the exam. Study situations should be as exam-like as possible, to prime the test-taker's system for the experience. By waiting to check the answers at the end, a psychological advantage will be formed, to decrease the stress factor.

Another fantastic reason for not cramming is the avoidance of confusion in concepts, especially when it comes to mathematics. 8-10 hours of study will become one hundred percent more effective if it is spread out over a week or at least several days, instead of doing it all in one sitting. Recognize that the human brain requires time in order to assimilate new material, so frequent breaks and a span of study time over several days will be much more beneficial.

Additionally, don't study right up until the point of the exam. Studying should stop a minimum of one hour before the exam begins. This allows the brain to rest and put things in their proper order. This will also provide the time to become as relaxed as possible when going into the examination room. The test-taker will also have time to eat well and eat sensibly. Know that the brain needs food as much as the rest of the body. With enough food and enough sleep, as well as a relaxed attitude, the body and the mind are primed for success.

Avoid any anxious classmates who are talking about the exam. These students only spread anxiety, and are not worth sharing the anxious sentimentalities.

Before the test also involves creating a positive attitude, so mental preparation should also be a point of concentration. There are many keys to creating a positive attitude. Should fears become rushing in, make a visualization of taking the exam, doing well, and seeing an A written on the paper. Write out a list of affirmations that will bring a feeling of confidence, such as "I am doing well in my English class," "I studied well and know my material," "I enjoy this class." Even if the affirmations aren't believed at first, it sends a positive message to the subconscious which will result in an alteration of the overall belief system, which is the system that creates reality.

If a sensation of panic begins, work with the fear and imagine the very worst! Work through the entire scenario of not passing the test, failing the entire course, and dropping out of school, followed by not getting a job, and pushing a shopping cart through the dark alley where you'll live. This will place things into perspective! Then, practice deep breathing and create a visualization of the opposite situation - achieving an "A" on the exam, passing the entire course, receiving the degree at a graduation ceremony.

On the day of the test, there are many things to be done to ensure the best results, as well as the most calm outlook. The following stages are suggested in order to maximize test-taking potential:

Begin the examination day with a moderate breakfast, and avoid any coffee or beverages with caffeine if the test taker is prone to jitters. Even people who are used to managing caffeine can feel jittery or light-headed when it is taken on a test day.
Attempt to do something that is relaxing before the examination begins. As last minute cramming clouds the mastering of overall concepts, it is better to use this time to create a calming outlook.
Be certain to arrive at the test location well in advance, in order to provide time to select a location that is away from doors, windows and other distractions, as well as giving enough time to relax before the test begins.
Keep away from anxiety generating classmates who will upset the sensation of stability and relaxation that is being attempted before the exam.
Should the waiting period before the exam begins cause anxiety, create a self-distraction by reading a light magazine or something else that is relaxing and simple.

During the exam itself, read the entire exam from beginning to end, and find out how much time should be allotted to each individual problem. Once writing the exam, should more time be taken for a problem, it should be abandoned, in order to begin another problem. If there is time at the end, the unfinished problem can always be returned to and completed.

Read the instructions very carefully - twice - so that unpleasant surprises won't follow during or after the exam has ended.

When writing the exam, pretend that the situation is actually simply the completion of homework within a library, or at home. This will assist in forming a relaxed atmosphere, and will allow the brain extra focus for the complex thinking function.

Begin the exam with all of the questions with which the most confidence is felt. This will build the confidence level regarding the entire exam and will begin a quality momentum. This will also create encouragement for trying the problems where uncertainty resides.

Going with the "gut instinct" is always the way to go when solving a problem. Second guessing should be avoided at all costs. Have confidence in the ability to do well.

For essay questions, create an outline in advance that will keep the mind organized and make certain that all of the points are remembered. For multiple choice, read every answer, even if the correct one has been spotted - a better one may exist.

Continue at a pace that is reasonable and not rushed, in order to be able to work carefully. Provide enough time to go over the answers at the end, to check for small errors that can be corrected.

Should a feeling of panic begin, breathe deeply, and think of the feeling of the body releasing sand through its pores. Visualize a calm, peaceful place, and include all of the sights, sounds and sensations of this image. Continue the deep breathing, and take a few minutes to continue this with closed eyes. When all is well again, return to the test.

If a "blanking" occurs for a certain question, skip it and move on to the next question. There will be time to return to the other question later. Get everything done that can be done, first, to guarantee all the grades that can be compiled, and to build all of the confidence possible. Then return to the weaker questions to build the marks from there.

Remember, one's own reality can be created, so as long as the belief is there, success will follow. And remember: anxiety can happen later, right now, there's an exam to be written!

After the examination is complete, whether there is a feeling for a good grade or a bad grade, don't dwell on the exam, and be certain to follow through on the reward that was promised...and enjoy it! Don't dwell on any mistakes that have been made, as there is nothing that can be done at this point anyway.

Additionally, don't begin to study for the next test right away. Do something relaxing for a while, and let the mind relax and prepare itself to begin absorbing information again.

From the results of the exam - both the grade and the entire experience, be certain to learn from what has gone on. Perfect studying habits and work some more on confidence in order to make the next examination experience even better than the last one.

Learn to avoid places where openings occurred for laziness, procrastination and day dreaming.

Use the time between this exam and the next one to better learn to relax, even learning to relax on cue, so that any anxiety can be controlled during the next exam. Learn how to relax the body. Slouch in your chair if that helps. Tighten and then relax all of the different muscle groups, one group at a time, beginning with the feet and then working all the way up to the neck and face. This will ultimately relax the muscles more than they were to begin with. Learn how to breathe deeply and comfortably, and focus on this breathing going in and out as a relaxing thought. With every exhale, repeat the word "relax."

As common as test anxiety is, it is very possible to overcome it. Make yourself one of the test-takers who overcome this frustrating hindrance.

Additional Bonus Material

Due to our efforts to try to keep this book to a manageable length, we've created a link that will give you access to all of your additional bonus material.

Please visit http://www.mometrix.com/bonus948/plumberjourney to access the information.